The
Bhopal Syndrome

Pesticides, Environment
and Health

DAVID WEIR
(Center for Investigative Reporting,
San Francisco)

D1355567

EARTHSCAN PUBLICATIONS LIMITED

LONDON

This edition first published in Great Britain 1988 by
Earthscan Publications Limited
3 Endsleigh Street, London WC1H 0DD

Published in the USA by Sierra Club Books

British Library Cataloguing in Publication Data

Weir, David
    The Bhopal syndrome.
    1. Pesticides. Environmental aspects
    I. Title
    632'.95042

ISBN 1-85383-011-9

Printed and bound in Great Britain by
Cox and Wyman Ltd, Reading, Berks

Front cover photograph: GLC
Back cover photograph: Associated Press Ltd

To
Alison,
Laila,
Sarah,
and Peter

———

Contents

APPENDICES

Foreword

In the early hours of December 3, 1984, the city of Bhopal, India, was converted into a gas chamber, creating a holocaust unprecedented in the annals of man-made industrial disasters. Bhopal was expensive in human lives, in environmental damage, and in economic and social costs. It was unnecessary and avoidable.

Tragically, Bhopal is being repeated, not just as explosions, infernos, and deadly clouds heard, felt, and seen, the world over, but as "mini-Bhopals"— smaller industrial accidents that occur with disturbing frequency in chemical plants in both developed and developing countries. Even more numerous and deadly are the "slow-motion Bhopals"—unseen and chronic poisoning from industrial pollution that causes irreversible pain, suffering, and death.

Also evident at Bhopal, and just as replicable, are the failures of both corporations and government bureaucracies to avert and control the incident; the scandalously inadequate emergency and medical responses; and the hopelessly inadequate post-event rehabilitation and compensation.

Like Auschwitz and Hiroshima, the catastrophe at Bhopal is a manifestation of something fundamentally wrong in our stewardship of the earth. Its ghosts will be with us forever, but we can and we must begin in earnest to stop what David Weir calls "the Bhopal Syndrome."

To do this, we need a new awareness, increased knowledge, and, most of all, the will to act locally and globally. We need actions that will go beyond mere "fire fighting" to directly confront and fundamentally rethink the present paradigms of development. We must move away from development strategies that are inherently violent, manipulative, and wasteful toward those that are humanized and ecologically sound.

We need to change the value systems of our industrial enterprises so that the health and safety of both people and the environment is paramount, superseding any technical or commercial considerations. We need to democratize technology to guarantee full disclosure about hazards. We need universal acceptance of the principles of "right to know" and "freedom of information" on health and safety issues.

Citizens' groups are beginning to create this global conscience and will to act. The movement started with alarm about the almost total failure of industry, governments, and international agencies to curb the proliferation of hazardous pesticides, like those produced at Bhopal.

A worldwide citizens' coalition of groups and individuals—the Pesticide Action Network (PAN)—is attempting to halt the spread and misuse of pesticides, particularly in Third World countries where the danger is greatest. PAN was formed in May 1982, following a conference in Penang, Malaysia, organized by the International Organization of Consumers Unions, and co-hosted by Sahabat Alam (Friends of the Earth) Malaysia. In 1985, PAN launched the international "Dirty Dozen" campaign, a public education effort coordinated by the San Francisco-based Pesticide Education and Action Project (PEAP), which targets 12 hazardous pesticides for global action. By late 1985, PAN participants were active through hundreds of groups in over 50 countries, emerging as a strong alternative mechanism through which concerned citizens can act.

The "No-More-Bhopals" network was inaugurated in 1985 at a Nairobi meeting on sustainable development organized by the Environment Liaison Centre (ELC) and the International Coalition for Development Action (ICDA). Scores of groups, from the Citizens Commission on Bhopal in the USA and the Bhopal Monitoring Resource Group in Japan to the Kerala Peoples' Science Centre in India, have produced waves of popular action.

This book by David Weir takes us on a journey through many continents and shows us the pervasiveness of "the Bhopal Syndrome." The incisive reporting and masterful writing makes compulsive and even thrilling reading. Claude Alvares, one of India's leading journalists, has added a stunning personal account.

The revelations will shock many. Let us hope it stirs us out of any complacency that can so quickly settle in once an event has faded from the mass media.

ANWAR FAZAL
Penang, Malaysia

Acknowledgments

For helping make the North American edition of this book a reality, I thank the staff members of the Center for Investigative Reporting (CIR), and of the regional office for Asia and the Pacific of the International Organization of Consumers Unions (IOCU) in Penang, Malaysia.

Parts of chapters one, five, and six first appeared as an article I coauthored with Martin Abraham for *Consumer Interpol Focus*, June 1985. Rani Advani contributed reporting help to chapter one; Daphne Wysham to chapters three and seven; Indra Tata to chapter seven; David Hathaway and Frederico Fullgraf to chapter eight; Cassie Tao to chapter nine; Nobuo Matsouka and Jun Ui to chapter eleven; and Connie Matthiessen to chapters twelve and fourteen. Claude Alvares wrote the Afterword, *A Walk Through Bhopal*.

Danny Moses, of Sierra Club Books, guided the writing of the Introduction and Conclusion, and edited the manuscript; Peggy Lauer provided a careful copyedit; Barbara Newcombe contributed the index.

Anwar Fazal of IOCU and Monica Moore of the Pesticide Education and Action Project in San Francisco suggested the idea for this book, and were supportive throughout its development. I was also helped by people in many countries, including Toshiaki Matsuzawa, Erna Witoelar, Magda Renner, Ivan Restrepo Fernandez, Antonio Thomen, Robert Northup, Alexander Bonilla, V. K. Pariji, Jens Huhn, Anke Schulz, Orlaith Kelly, Thierry Lemasquier, Colin Moorcraft, José Vargas, Dorothy Myers, Gretta Goldenman, David Cobb, Don Lewis, Molly Coye, Sandra Marquardt, Richard Wiles, Larry Everest, Howard Saunders, Gerardo Velazquez, Jim Kettler, Dick Johnson, Dorian Kinder, Judy Donald, John Hunting, Sally K. Fairfax, Ann Roberts, Terry Crowley, Barbara Newcombe, Henry Weinstein, Lauren Volpini, Andrew Porterfield, and some people whose identities must remain confidential.

Organizations that helped us include the Ruth Mott Fund (Flint); the Beldon Fund (Washington, D.C.); the CS Fund (Freestone); the Strong Center (Berkeley); the W. Alton Jones Foundation (Charlottesville); the Jessie Smith Noyes Foundation (New

York); the Mary Reynolds Babcock Foundation (Winston-Salem); the People's Research Institute on Energy and Environment (Tokyo); the Bhopal Disaster Monitoring Group (Tokyo); the Consumers Union of Japan; Jishu Koza (Tokyo); citizens and workers in Kitakyushu; the Consumers' Foundation R.O.C. (Taipei); the Indonesian Environmental Forum (Jakarta); the Indonesian Network Against the Misuse of Pesticides (Jakarta); the Institute for Legal Aid (Jakarta), Youth Farmers for Rural Development (Indonesia); Sahabat Alam (Malaysia); Consumers Association of Penang; Acao Democratica Feminina Gaucha (Porto Alegre); the Consumer Education and Research Centre (Ahmedabad); the Consumer Education Centre (Hyderabad); Transnationals Information Exchange (Amsterdam); the International Federation of Plantation and Agricultural Workers (Geneva); Oxfam (U.K.); the German Marshall Fund (Washington, D.C.); the Washington Research Institute (San Francisco); the Highlander Center (New Market, Tennessee); the National Coalition Against the Misuse of Pesticides (Washington, D.C.); the Labor Institute (New York); the Workers Policy Project (New York); the Data Center (Oakland); and the Pesticide Action Network, International.

I thank all of these people and groups for their help; but the responsibility for this book's shortcomings is my own.

INTRODUCTION

Our Need
to Know

There are things we would rather not know about; yet we must know. What happened at Bhopal, India, in December 1984 is one of those things.

If the Bhopal disaster had been a once-in-a-life-time event, it would be possible to ignore it, forget it, or assign it to that curious category called information of merely academic interest. But the Union Carbide pesticide plant that sent poison gases pouring into the air over Bhopal was much closer to all of our lives than we commonly realize. It was part of a world-wide food production system that affects nearly every person on earth. The potential for similar accidents occurring elsewhere is high. According to the Congressional Research Service, about 75 percent of

the population of the United States lives "in proximity to a chemical plant."[1] Many of these factories
manufacture pesticides; the United States is the top
producer—and user—of pesticides in the world.

Ours is an age of environmental pollution. To survive as a viable species, we need to find solutions to
many problems. The basic elements of life on earth—
air, water, soil—are increasingly contaminated or
otherwise compromised for future use. The tough and
resilient web of plants and other animals, which sus-
tains our existence, is now weakening strand by
strand, and is in danger of eventually disappearing
beneath us.

Above us, chemical pollutants from our everyday
industrial and consumer activities have been blamed
for opening a hole in the ozone layer of the earth's
atmosphere. This hole is apparently growing—along
with implications that are terrifying to consider, such
as the possibility of millions of additional cases of
skin cancer, deaths of many aquatic creatures at the
larva stage, adverse effects on crops and other plants,
and global climatic changes that would curtail human
activity on earth.[2]

There are additional problems. Small amounts of
many different chemicals (as well as heavy metals) are
embedded in the tissues of many living creatures, especially humans. We do not know what the health
effects of these substances may be, but a frightening

new line of research indicates that the all-important yet poorly understood human immune system may be weakened due to exposure to chemical pollutants.[3] To date, the work is preliminary and can be considered only suggestive, but again, the implications are frightening. A weakened immune response would leave us vulnerable to the many disease-causing viruses and bacteria that continue to evolve and mutate along with all life on the planet.

Our knowledge of these and other aspects of the ecological crisis we face, though fragmentary, far outstrips our current ability to do anything about them. A feeling of helplessness weighs down those who try to confront the terrifying issues of our time. Nothing symbolizes our apparent impotence better than the specter of the superpowers, armed with nuclear weapons and threatening to end the human debate, once and for all, over which political economy is the better scheme.

Distrustful of one another, posturing on a global stage, the leaders of the United States and the Soviet Union seem to be incapable of finding the solutions we so desperately seek. Together they comprise two factions of a global power structure that excludes the vast majority of the world's people. Most human beings lead marginal lives under oppressive forms of social organization. Many people, rich or poor, have diminishing hopes for the future.

The bounty of industrialization, especially under the "free enterprise" faction, has created a slender slice of the global population that is well-housed, clothed, fed, and educated. With education has come a broadening awareness among this group that humans are only one kind of animal in an intricately balanced ecosystem. This awareness was already present in the cultures of many indigenous people around the world. But these cultures, along with the tropical rain forests that sustain them, are now in danger of disappearing. What distinguishes the present moment is that large numbers of people all over the world understand that collectively—rich and poor—humans as a species represent a dictatorship of life on earth.

Like all dictatorships, this one is corrupt. The human species uses its superior physical and mental traits to plunder the resource wealth of the globe, with little concern for the effects on other life. Not content with mere comfort and health, we in the developed countries are piling up a huge surplus, squeezed out of rocks, rivers, and forests, leaving only deserts in return. Deserts, at least the kind created by human activity, are exhausted stretches of earth, eroded of life and sustenance. They can be reclaimed only by intensive applications of resources better preserved for other needs.

Besides desertification, we are as a species actively engaged in deforestation, soil erosion, water depletion, and overpopulation of the planet. We seem suspended in time, refusing to acknowledge the gravity of the ecological crisis, slow to commence the search for renewable forms of energy, the "soft" path to a future based on logic, rather than on "the market."

Dictatorships of any type, however, do not last forever. The seeds that grow to destroy these oppressive systems can be easily seen in our present situation. Other living creatures are capable of surviving whatever we eventually do to the planet, succeeding us, and establishing a new order. The only question is whether the self-extinction of the current dictator can be avoided.

Since the Bhopal pesticide disaster, the world has been shaken by additional industrial accidents of major ecological significance. The worst occurred in the spring of 1986, in Chernobyl, USSR. A nuclear reactor exploded, spewing radioactivity into the atmosphere, which then carried it to every continent, contaminating food supplies all over the world.[4]

Then came what Europeans call "ChernoBasel" or "Sandoz-Bhopal" in the fall of 1986. A fire at a chemical warehouse owned by the multinational Sandoz, Inc., in Basel, Switzerland, led to the dumping of massive amounts of toxic substances, including 66,000 pounds of pesticides, into the Rhine River, one of the most important waterways in the world. Ciba-Geigy, a multinational neighbor of Sandoz, also released a smaller load of toxic pesticides into the river during the same period.[5]

Public concern over these tragic events, like the one at Bhopal, eventually passed, and by now they are more or less forgotten by most people. Taken one by one, they will eventually be added to history's scroll as footnotes or, less likely, as major entries. These judgments are normally determined by those who come afterward.

But there may not be time to wait for history's judgment. Those of us not yet numbed by the bewildering array of social, political, and economic problems our societies face must initiate public debate that will lead to a series of basic environmental decisions. Based on the evidence available, and our interpretations of that evidence—and based also on our instincts, which are not entirely foolish—we have to make choices that will immediately affect our future and the future of our children. One choice is to work toward the elimination of chemical pesticides from agriculture.

The word for pesticide translates as "medicines for food" in many languages, including those spoken by the people who lived along the walls of the Union Carbide factory in Bhopal. Pesticides once were called "wonder drugs" in the United States, as well. They stimulated crop yields undreamt of by our ancestor farmers, and today, it is a common belief that we cannot grow food without them. A more accurate statement would be that we could grow all our food without pesticides, but more of us would have to be involved in the process. *The Planter*, a trade magazine of the Incorporated Society of Planters in Malaysia, discussed this in an editorial in its June 1985 issue:

"The crux of the matter is that these chemicals are needed in the world . . . if we are to maintain some degree of control over some of the biological antagonists against which we have to fight. This now, and on into the future. No doubt, the earth and man have existed for a long time without such chemicals, which only came on the scene in the middle of the century. But, in that time, they have greatly altered our capacity to protect our crops. Far greater yields can be achieved with them, if they are properly used, with far less work. In this respect, they are an essential part of modern life. So the question appears to be, do we want to continue to live as at present, or to return to drudgery?"[6]

As consumers, we react with dismay to the news that small amounts of pesticides routinely find their way into our food supply. If we knew more about the way in which chemicals interact inside our bodies, we might be even more alarmed. But whatever our re-action, it is that of a relatively passive participant in the food production system—not somebody actively engaged in the process by which our food arrives at market.

In economic terms, the primary role for most of us in today's world is that of consumer, and we seem to be frantically consuming an unprecedented diversity of products, compliments of the global production frenzy that characterizes our age.

The evidence that today's food production system is global is all around us. Americans drink coffee from Brazil and Kenya, and eat fruits and vegetables from Mexico and Peru, meat from Central America and New Zealand, bananas from the Philippines (all, perhaps, on a teakwood table stripped from rain forests in Indonesia). More than ever before, we are the end consumers of a truly global food system, tied inextricably to farm workers and their families in faraway places—people who are getting sick and dying while planting and harvesting our foods. For, make no mistake about it, pesticides kill. Hundreds of thousands of people are poisoned annually, over ten thousand fatally. These are the relatively conservative estimates of the World Health Organization. Nobody knows how high the actual toll rises each year.[7] In the Third World, where most of the poisonings occur, the long-term health risks are not even being studied. In the United States, where recent immigrants harvest much of our domestic produce, the United Farm Workers Union has identified pesticide risks both to workers and consumers as the top issue on its agenda.

During the past decade, pesticides have become embedded in agricultural systems all over the world, contributing to the ever more capital-intensive food production system we have come to depend on. Following those expanding pesticide markets, the multinational chemical companies—as well as a host of local imitators—have set up manufacturing and formulating facilities in practically every populated corner of the world. The manufacture and use—and, most importantly, the *misuse*—of pesticides is now ubiquitous. The effects of the global pesticide trade thus touch us all.

There is, however, another way to grow our food. And it is within our power—as consumers who sustain the present system—to demand it. The small, casual choices we make in the supermarket each week add up to the driving force that makes or breaks seemingly mighty industries like the global agrochemical industry. Our society has the scientific knowledge to grow food for everyone without using deadly chemicals. But transforming the institutions that currently comprise the food production system will be difficult. There may indeed be some return to "drudgery," in the form of greater use of human labor and less use of "wonder drugs."

But for those who are without jobs and are unable to save their children from suffering from malnutrition and disease—and there are many—such work can be an opportunity, not drudgery. And even in the developed countries, where social and educational systems seem to remove meaning and hope from life rather than encourage our abilities and harness our idealism, there is a large potential agricultural work force. These human resources need to be tapped in fair, just ways—a departure from the current methods of exploitation that characterize most farm working conditions.

Although the choice we as consumers must make about pesticides is just one of the difficult decisions we face, it is at least a manageable one. Perhaps in finding our way to a nonchemical food production system, we will develop the organizational tools necessary to work on some of the other unfinished business.

The first step, however, is getting the facts. That is what this book is all about.

Notes

1. James McCarthy, Congressional Research Service Report (Washington, D.C.: Feb. 22, 1985).

2. Mark Crawford, "United States Floats Proposal to Help Prevent Global Ozone Depletion," *Science*, vol. 234, Nov. 21, 1986, p. 928.

3. "Study of Pesticide Aldicarb Suggests Effect on Humans," *Wall Street Journal*, Oct. 28, 1986; see also Leon John Olsen, et. al., "Aldicarb Immune Suppression in

Mice: An Inverse Dose-Response to Parts per Billion Levels in Drinking Water," (draft), University of Wisconsin Center for Environmental Toxicology, June 1985; see also George P. Casale, Steven D. Cohen, and Richard A. DiCapua, "Parathion-Induced Suppression of Humoral Immunity in Inbred Mice," *Toxicology Letters*, vol. 23, 1984, pp. 239–47; see also "Already Controversial Dioxin Document Under Internal EPA Review," *Pesticide & Toxic Chemical News*, Oct. 1, 1986, p. 14.

4. Associated Press, "Europeans Check the Levels of Radioactivity," *New York Times*, May 2, 1986.

5. NCAMP's Technical Report, vol. 1, no. 7, *News Bulletin of the National Coalition Against the Misuse of Pesticides* (Washington, D.C.: Nov. 1986).

6. "The pipe-dreams of PAN?" *The Planter* (Kuala Lampur, Malaysia), June 1985, pp. 245–48.

7. David Bull, *A Growing Problem: Pesticides and the Third World Poor* (Oxford, UK: Oxfam, 1982).

PART
ONE

The
Tragedy

ONE

—

Running Toward Bhopal

Central India. December 2, 1984: In a village south of the city of Bhopal, a small boy named Javed sees an airplane flying overhead. Entranced, he begins to run, following in the direction of the plane, running toward Bhopal. . . .

In the dead of night, a 12-year-old boy named Ilayas and his friend Babla arrive at the Bhopal railway station, "ticketless," having hopped a train from the south after a scolding by Ilayas's father. . . .

In a shantytown nearby, not far from the mammoth pesticide factory owned by Union Carbide Corporation, another young boy named Suresh is awakened by his parents. His mother places the house key on a string around his neck, grabs his hand, and pulls him outside to flee. Somehow, however, the family becomes separated in the human stampede that envelops Bhopal's streets. . . .

Javed, Ilayas, and Suresh were all survivors of the worst industrial accident in history—those who lived to tell about it. Afterward, they were placed in an orphanage in Bhopal.[1] Babla died on the railway platform and Suresh's parents were apparently among the many unidentified dead.

Very early on the morning of December 3, 1984, a violent chemical reaction occurred in a large storage tank at the Union Carbide factory. A large amount, perhaps 40 tons, of methyl isocyanate (MIC)—a chemical so highly reactive that a trace contaminant can set off a chain reaction—escaped from the tank into the cool winter's night air. A yellowish-white fog, an aerosol of uncertain chemical composition, spread over the sleeping city of 800,000.* The mist hovered close to the ground—MIC is heavier than air—and blanketed the slums of Bhopal. Hundreds of thousands of residents were rousted from their sleep, coughing and vomiting and wheezing. Their eyes burned and watered; many would soon be at least temporarily blinded. Most of those fortunate enough

---

* The exact chemical composition of the gas is a matter of continuing controversy. See the Afterword, page 159, for a discussion of this controversy and its impact on the health problems of the gas-leak victims. For purposes of simplicity, the killer gas is hereafter identified as MIC.

to have lived on upper floors or inside well-sealed buildings were spared. The rest, however, opened their doors onto the largest unplanned human exodus of the industrial age. Those able to board a bicycle, moped, bullock car, bus, or vehicle of any kind did. But for most of the poor, their feet were the only form of transport available. Many dropped along the way, gasping for breath, choking on their own vomit and, finally, drowning in their own fluids. Families were separated; whole groups were wiped out at a time. Those strong enough to keep going ran 3, 6, up to 12 miles before they stopped. Most ran until they dropped.[2]

By now, the human tragedy that occurred that night has dimmed in most people's memories, eclipsed by the continuing drama of other news events, large and small, of the intervening months and years. In most ways, life goes on. Bhopal has settled back to a kind of "normalcy," though thousands of the victims of the gas leak still suffer symptoms of serious, long-term health damage, including incurable problems with breathing, sleeping, and digesting food.[3] A large number of women are reported to be suffering from gynecological disorders, and numerous spontaneous abortions and birth defects have been recorded.[4] Meanwhile, the various principals to the tragedy—Union Carbide Corporation, headquartered in Danbury, Connecticut; its Indian subsidiary,

Union Carbide India Ltd. (UCIL), which operated the Bhopal plant; the government of India; and the legal representatives of the dead and injured—are suing each other in court. Billions of dollars are at stake, but Union Carbide's management has steadfastly insisted that there will be no adverse financial consequences from the various lawsuits.

The Bhopal accident has remained more of a mystery than some other industrial disasters. Many basic questions about the catastrophe have yet to be resolved, with no serious prospect that either the Indian government or Union Carbide will provide answers. In the wake of the accident, public concern about the safety of pesticide technology has grown, just at the time when formulation and manufacturing plants have spread to practically every corner of the planet. Industry sources indicate that there are over 250 major producing companies in the world now, many with multiple plants in multiple countries.[5] As formerly remote parts of the Third World are penetrated by the inexorable spread of pesticide technology, local formulation plants are springing up, sometimes with lower standards and weaker controls than those of the multinational corporations that dominate the business on a global basis. This increases the potential for future disasters.

This book investigates a number of questions that

seem particularly relevant given the scope of the Bhopal accident. What are conditions like at other pesticide plants in underdeveloped countries? Are extremely hazardous substances like MIC in common use? Do any of the plants have a history of accidents or leaks that might foreshadow future problems? Are people living near these plants? If so, are there adequate evacuation plans to keep casualties to a minimum?

The answers to these and related concerns provide a framework for raising the most important question of all: *Could a Bhopal tragedy happen again?* Since December 1984, executives, workers, regulators, insurers, reporters, and citizens all over the world have been asking that question.

This book intends to provide some of the evidence needed so that the public debate on this issue might be an informed one. For the most part, it focuses on pesticide plants, like that at Bhopal, though there are many types of hazardous technologies. It also places the conditions at these plants in the context of Third World food production, since that is the stated rationale for the rapid proliferation of pesticide plants around the world.

First, then, we have to engage in a quick survey of the global pesticide business by examining the products, the companies, and the momentum behind its remarkable rate of growth.

Notes

1. Personal interviews with Rani Advani, Bhopal, India, Dec. 21, 1984.

2. Numerous press accounts.

3. Steven R. Weisman, "Disabling and Incurable Ailments Still Afflict Thousands in Bhopal," *New York Times*, Mar. 3, 1985.

4. Padma Prakash, "Bhopal Study Reveals High Incidence of Pelvic Disease," *Sunday Observer* (Bombay), Mar. 17, 1985; see also "Six Deformed Babies Linked to Bhopal Gas Leak," *New York Times*, July 16, 1985.

5. Confidential pesticide industry report.

TWO

—

The Global Pesticide Industry

Pesticides are big business. Approximately $13 billion worth of pesticides were sold around the world in 1983—the result of a quarter century of sustained growth in sales averaging 12.5 percent per year.[1] This explosion in the sales of man-made chemicals is equaled by very few products in any other industry.

One way to classify pesticides is by what they are intended to kill; and so four of the basic categories are insecticides, herbicides, fungicides, and rodenticides. Another method of identification is by chemical class, such as organochlorines, organophosphates, carbamates, and so on. The best known of all pesticides, DDT, is an organochlorine insecticide that was first synthesized over a century ago, but which became a commercial bug-killer only at the end of World War II.[2] Another of the first generation of insecticides, parathion, was originally devised as an agent for chemical warfare by Nazi scientists.[3] At the end of the war a burgeoning chemical industry was without a market for its wartime products. But that soon changed. Before long, the tools of war between men were being employed in the war against pests.

Since the dawn of agriculture, people have been trying to improve their terms of competition with other creatures for food and fiber. The most formidable "pests" are insects, the largest single group of animals on the planet. Only a very small portion (less than 1 percent) of the 1,250,000 insect species known to science are pests,[4] but they effectively destroy up to a third of our crops every year.[5]

It is against this army of pests that synthetic organic chemicals are employed. And, despite 40 years of intense chemical warfare against our natural competitors, the record is mixed. Although "miracle" crop yields were achieved in the early years, thanks to pesticides, the insects have reasserted themselves, with a result that could best be termed a standoff. The overall percentage of each year's crop lost to pests has remained the same, despite the use of pesticides. And, all over the world, insects are developing resistance to the weapons in our chemical arsenal, with a

16-fold increase in the number of arthropod pests (insects, mites, and ticks) resistant to one or more pesticides between 1955 and 1980.[6]

In an attempt to keep ahead of the pests, farmers in many parts of the world are using greater and greater amounts of increasingly toxic pesticides, thereby helping fuel the global explosion in chemical sales. In the process, however, they are triggering more and more resistance in pests, creating what has been called the "pesticide treadmill."

One result of this escalating reliance on pesticides is that growing food has become much more expensive. But the profits in the food production business are not necessarily going to the growers. On proprietary products, the pesticide manufacturers commonly make profits in the range of 17 to 25 percent.[7] Most of these profits go to the multinationals, which have the research, manufacturing, and marketing expertise to develop and distribute new products. Typically, pesticides are produced by a single division of a large corporate enterprise—Union Carbide, for instance, derives only about 3.6 percent of its sales worldwide from pesticides.[8] About three dozen companies control over 90 percent of the world trade in pesticides, with the top ten accounting for over 50 percent of that total. Three companies alone—Bayer of West Germany, Ciba-Geigy of Switzerland, and Monsanto of the United States—controlled 25 percent of the world pesticide market, or over $4 billion in sales, in 1983.[9]

This highly concentrated industry spans the globe, with networks of subsidiaries and affiliates throughout not only the industrialized countries but most of the underdeveloped countries as well. The Agricultural Chemicals Division of Bayer, for example, sells pesticides through 50 overseas subsidiaries in over 100 countries.[10]

During the past decade, as the agricultural pesticide markets in the developed countries approached saturation, the multinationals have turned more and more to exports, particularly to the booming markets of the Third World.[11] In Africa, for instance, pesticide use was projected to have *quintupled* during the past ten years.[12]

By moving into the Third World, the multinationals have encountered conditions quite unlike those at home. Malnutrition, illiteracy, poverty, and short life spans are the norm. Economic development, the priority of virtually every government in the world, lags far behind that of the United States, Europe, and Japan. By 1974, a decade before the Bhopal tragedy, for example, Union Carbide was marketing its products in 125 countries, 75 of which had smaller economies than the corporation.[13] Holding such an advantage in size and money over many host governments gives the companies a great deal of leverage. Though they seldom comment publicly about their relationship with Third World government officials, internal company documents reveal intense competition among firms to influence policymakers in host countries. For example, a 1978 "travel report" issued inside Hoechst, the West German chemical giant that ranks as the world's seventh-largest pesticide seller, refers to that company's close relations with the relevant officials setting pesticide policy for Egypt. The report warns, however, that one of its competitors, Ciba-Geigy, also "has excellent relations with the Ministry of Agriculture."[14]

In a similar report on pre-Sandinista Nicaragua, in 1978, Hoechst officials complain of "strong pressure exercised by Bayer on the authorities to use Bayleton [a Bayer pesticide]," including "gifts" and an "invitation" to travel to company headquarters in Germany. "Hoechst or our distribution company, Cateran, has no means of countering this growing pressure from Bayer," the report warns, although it recommends new "investment" in "gifts," in order "to counter the pressure from Bayer."[15]

While keeping track of the competition, regional officials of the multinationals are responsible for charting the course to long-term profits in their area. An internal "strategic analysis" of the Latin America market for 1976 to 1985 by officials from BASF, another West German multinational, which ranks ninth in the world pesticide trade, puts it this way: "It is a general objective to keep pace with the overall growth rates of the Latin American market or to exceed these in order to achieve maximum exploitation of the existing profit possibilities."[16]

As part of this relentless search for profits, the chemical giants have erected manufacturing and formulation facilities throughout the Third World, with dozens of new plants opening every year. One of these factories was the now infamous Union Carbide plant at Bhopal. Its story shares important similarities with all the rest.

Notes

1. Wood, Mackenzie & Co. Agrochemical Service, 1984.
2. George M. Woodwell, "Broken Eggshells," *Science 84*, Nov. 1984.
3. Lewis Regenstein, *America the Poisoned* (Washington, D.C.: Acropolis Books Ltd., 1982), pp. 102–5.
4. R. H. Hall, "A New Approach to Pest Control in Canada," Canadian Environmental Advisory Council Report, no. 10, July 1981.
5. T. J. Sheets and D. Pimentel, eds., *Pesticides: Their*

Contemporary Roles in Agriculture, Health and the Environment (Clifton, N.J.: Humana Press), pp. 97–149.

6. Donald Dahlstein, *Environment*, vol. 25, no. 10, Dec. 1983; see also: G. P. Georghiou and T. Saito, eds., *Pest Resistance to Pesticides* (New York: Plenum Press, 1983), pp. 1–46.

7. Frost & Sullivan, Inc.

8. Ibid.

9. Wood, Mackenzie & Co., 1984.

10. Ibid.

11. Predicasts, Inc., quoted in *Futurist*, April 1984; see also Jeanie Ayers, "Pesticide Industry Overview," *Chemical Economics Newsletter*, Jan.–Feb. 1978.

12. "Better Regulation of Pesticide Exports and Pesticide Residues in Imported Food is Essential" (Washington, D.C.: U.S. GAO Report no. CED-79-43, June 22, 1979).

13. "Union Carbide: A Study in Corporate Power and the Case for Union Power," Oil, Chemical and Atomic Workers International Union, June 1974.

14. Dr. Mildner, "Pflanzenschutz: Reisebericht Ägypten," report for Hoechst Aktiengesellschaft, FRG, Nov. 13, 1978.

15. Jenspeter Meyer and Dr. Dieterich, "Besuch Nicaragua," report for Hoechst Aktiengesellschaft, FRG, Nov. 20, 1978.

16. "Sparte Lateinamerika: Strategische Analyse, 1976–1985," BASF, FRG.

THREE

A Disaster
Waiting
to Happen

Until the sale in 1986 of its consumer products division, Union Carbide's best-known consumer product was the *Eveready* battery. In India, as in the U.S. and many other countries, *Eveready* is a household name, partly due to the company's long-term promotional efforts. Twenty years ago, over 100 Union Carbide vans roamed through the villages in India's countryside with exhibits and recordings and a salesman who popped out to exclaim: "I'm the *Eveready* man."[1]

But the company's history in India goes back much farther than that, to the early 20th century, when the subcontinent was still part of the British Empire. It was the India of Rudyard Kipling and

29

E. M. Forster that Union Carbide first entered in 1905.[2] By the time Union Carbide India Ltd. (UCIL), the multinational's local subsidiary, decided to open a small pesticide formulating facility at Bhopal six decades later, the differences between corporate operating practices at the company's 13 Indian plants and those in the United States were well established. *Business Week* magazine reported in 1964 that UCIL's Camperdown battery works in Calcutta was "crustworn with age, dimly lit, poorly laid out for an assembly operation—and bursting at the seams." Although the magazine blamed the regulatory policies of India's "socialistic" government for the state of UCIL's battery plant, it also reported that a brand new Union Carbide chemical factory near Bombay, strongly supported by the government, "may not be up to U.S. plant standards."[3]

The relative responsibilities of Union Carbide and the Indian government for the state of the company's manufacturing facilities there would continue to be the cause of controversy two decades later. Unlike governments in smaller, weaker nations, the New Delhi government could insist on certain concessions from large corporations wishing to do business in India. It is a measure of Union Carbide's influence, however, that despite Indian law limiting foreign ownership of corporations to 40 percent, the U.S. parent company was allowed to retain majority ownership

(50.9 percent) of UCIL, because it was considered a "high technology" enterprise.[4]

In any event, it was partly in response to government incentives that Union Carbide launched its Bhopal operation on a five-acre plot of land in 1969. The company was reportedly allowed to build the plant on government land at an annual rent of less than $40 an acre, including taxes, as part of a scheme to attract industry to Bhopal, the capital city of Madhya Pradesh, the largest and one of the most economically depressed states in the nation.[5]

The initial $1 million investment set up a formulating plant to import, dilute, package, and ship *Sevin*, Union Carbide's most important pesticide. Hoping to take advantage of a growing market in India for carbamates like *Sevin*, Union Carbide expanded the plant several times during the 1970s and finally opened a full-fledged MIC production component in 1980.[6] By 1984, the plant was a $25 million manufacturing facility sprawled over 80 acres of Bhopal.

MIC is only one of many "intermediates" used in pesticide production. Union Carbide had alternative processes, not requiring MIC, for producing *Sevin* (also known by its chemical name, carbaryl), but they were more expensive and produced a larger volume of troublesome waste material.[7]

MIC is a particularly dangerous chemical. It is a

little lighter than water but twice as heavy as air, meaning that when it escapes into the atmosphere it remains close to the ground. It is highly inflammable and volatile, boiling at a temperature of 39.1 degrees C (102.4 degrees F). Most critically, MIC has the ability to react with many substances: water, acids, metals, and the small deposits of corrosive materials that accumulate in pipes, tanks, and valves. These reactions are extremely vigorous and heat-producing. Furthermore, given the presence of a catalyst, which could be a tiny bit of corrosion, MIC reacts with itself, quickly developing into a violent chain reaction. According to the trade journal *Chemical & Engineering News* (*C&EN*) in March 1985: "It's not hard to imagine situations in which such trace contamination could occur, despite best efforts at prevention."[8]

Moreover, one of MIC's components, phosgene (a nerve gas used in World War I), contains chlorine, which is able to attack most stainless steel alloys and produce the kinds of catalysts that could set off a runaway MIC reaction.[9] MIC is therefore difficult and dangerous to store, even in stainless steel tanks like those used at the Bhopal plant.

"Carbide was well aware of [MIC]'s hazardous nature," reported *C&EN*, yet the company went ahead with the design for the Bhopal plant, which included three huge storage tanks for MIC, each with a capacity of 15,000 gallons (45 metric tons). In order to prevent an uncontrolled MIC reaction, the parent company devised a series of safety measures and systems to contain the chemical, or at least slow down or stop a chemical reaction once it began. The company rules were aimed at preventing MIC from escaping, getting overheated, or being contaminated.[10]

Although these safety systems are automated with a state-of-the-art computer system at Union Carbide's plant in Institute, West Virginia, which also uses MIC in the production of *Sevin* and the other carbamate pesticides, many of the controls at the Bhopal plant were manually operated. Critics charge that this represented a "double standard," a characterization Union Carbide denies. The company says it had specified the design standards for the Bhopal factory, but the actual construction was done by its Indian subsidiary, UCIL, which used local equipment and material. Industry publications say that the Indian government required manual controls wherever possible.[11]

These kinds of requirements for local "input" into projects, as well as a degree of local control of ownership of foreign companies operating inside their borders, seem to be necessary strategies for Third World countries hoping to share the economic benefits of foreign investment. Furthermore, a labor-

intensive approach is the only logical way to industrialize in countries with many people who need jobs.
But in this context, and with the benefit of hindsight,
the decision to set up a plant using the extremely
hazardous MIC was a questionable one. Apparently
due to an anticipated doubling of demand for *Sevin*,
Union Carbide went ahead.

It soon became apparent that a serious error had
been made. A financial crisis hit Indian farmers in the
1980s and they began buying cheaper pesticides than
what Union Carbide could offer. The 1980s also
brought a new class of pesticides onto the market—
synthetic pyrethroids, which are not only less toxic
but can be used in much smaller quantities than *Sevin*.

In 1981, three of Union Carbide's competitors,
including the U.K.-based Imperial Chemical Industries (ICI), were licensed to sell synthetic pyrethroids
in India. Concerned that it would lose its market for
Sevin, Union Carbide began to work behind the
scenes.

According to an official directly involved, one
of the company's regional managers gathered documents from the U.S. Environmental Protection Agency about possible adverse health effects (including
cancer) from synthetic pyrethroids. He then dissem-
inated these documents to research scientists, hoping
they would contact the Indian Department of Agri-

culture and oppose the licensing of pyrethroids. Everything worked according to Union Carbide's plan until the company was asked to provide expert witnesses to testify against the pyrethroids at a government hearing. The company refused, apparently not wanting to anger its competitors by going public with its opposition.[12] The licenses for Union Carbide's competitors were granted, and the market for *Sevin* began to shrink.

The Bhopal plant, which had been designed for a production capacity of 5,000 tons of pesticides a year, never even came close to its target. In 1982, it produced 2,308 tons; in 1983, only 1,647 tons. In 1984, it reportedly produced less than 1,000 tons—a fraction of its capacity. The plant barely broke even in 1981; then it began losing money, running nearly $4 million in the red in 1984.[13] "The market never exceeded half our capacity," V. P. Gokhale, UCIL's managing director, told the *New York Times*, adding that "if the market would have expanded, we would have had a gold mine."[14]

Instead, Union Carbide had a white elephant right in the middle of a city that had more than doubled in population since the company had set up shop. Part of Bhopal's attraction to Union Carbide is its location at the center of India, with good rail connections to all parts of the country. The company was allowed to locate just two miles from the Bhopal railway station, which was convenient for shipping, but disastrous, it would turn out, for travelers on the night of December 2, 1984. For years, the plant has been ringed with shantytowns, mostly populated by squatters, who are part of the mass migration from countryside to city taking place all over the Third World. Although Union Carbide has claimed the squatters arrived after it did, old maps of Bhopal indicate otherwise. All three of the worst-affected communities in the disaster apparently existed before the Union Carbide plant opened.[15]

M. N. Buch, then a planning director for the state, remembers that in 1975, when Union Carbide received central government approval to manufacture MIC at Bhopal, he ordered the company to relocate to Bhopal's "obnoxious industries zone" 15 miles away. The Bhopal development plan, which had been devised earlier that year, called for the Union Carbide site to be converted to housing and commercial use. But Buch soon left Bhopal for another job and Union Carbide stayed where it was.[16]

The Bhopal plant, producing and storing an extremely unstable compound like MIC, lacking the state-of-the-art leak-detection equipment, and surrounded by teeming slums on every side, was a disaster waiting to happen.

Notes

1. "Thriving Under Fetters," *Business Week*, Nov. 7, 1964.

2. "The Trade Union Report on Bhopal," International Federation of Chemical, Energy and General Workers' Unions, Geneva, Switzerland, July 1985, p. 6.

3. "Thriving Under Fetters."

4. "Trade Union Report on Bhopal," p. 6.

5. "Pesticide Plant Started as a Showpiece but Ran into Troubles," *New York Times*, Feb. 3, 1985.

6. Ibid.

7. J. C. Forman, "Bhopal in Perspective," *Chemical Engineering Progress*, May 1985.

8. Ward Worthy, "Methyl Isocyanate: The Chemistry of a Hazard," *Chemical & Engineering News*, Feb. 11, 1985.

9. Ibid.

10. Worthy, "Methyl Isocyanate."

11. Forman, "Bhopal in Perspective."

12. Confidential interview with the author.

13. "Showpiece Ran into Troubles."

14. Ibid.

15. M. N. Buch, personal interview with author, Mar. 20, 1985.

16. Ibid.

FOUR

A Night of Terror

Union Carbide's Bhopal plant was, in the words of M. N. Buch, "a spanking new plant," rising above the narrow, crowded streets and ramshackle hovels of the city's poorer districts. But danger signals sounded everywhere, had anyone in authority been willing to listen.

Late in 1981, during the plant's first year of operation, one worker was killed and three others injured from exposure to phosgene, a gas used in the manufacture of MIC. The following year, local journalist Raj Kumar Keswani published a series of three exposés in a weekly newspaper about the inadequate safety standards at the plant, and the danger they represented to the citizens of Bhopal. Four days after

Keswani's second article, 18 workers were exposed to a mixture of three chemicals, including MIC, in a leak at the plant. Keswani says he then wrote directly to Arjun Singh, the chief minister of the state, warning him of the dangers from the plant, but received no answer.

Undeterred, Keswani continued his campaign to alert officials and the public. In June 1984, less than six months before the disaster, he published another story in *Jansatta*, a Hindi publication of the nation's largest newspaper chain, *Indian Express*, warning of the impending danger at Bhopal. If a massive gas leak should occur, Keswani wrote, "there will not even be a solitary witness to testify to what took place."[1]

Two years before, in May 1982, a three-member safety team from the Union Carbide headquarters in the United States visited the UCIL plant, and submitted a revealing report on the safety dangers of the MIC section. The report, which remained confidential inside Union Carbide until Keswani unearthed it in 1984, indicated a "serious potential for sizeable releases of toxic materials" in the MIC unit "either due to equipment failure, operating problems, or maintenance problems." The report questioned "the adequacy of the tank relief valve to relieve a runaway reaction" of MIC; noted that there was "frequent pressure gauge failure in all units"; and raised questions about workers employed "without having gained suf-

ficient understanding of safe operating procedures."
The report recommended various changes to reduce
the danger at the plant; there is no evidence the recommendations were ever implemented.[2] Workers and
union officials at the plant soon began issuing warnings of their own, talking with reporters and, according to various press accounts, putting up posters
in the surrounding communities to alert people to the
danger.

But authorities failed to heed these warnings. It
took the disaster to prove that the doomsayers were
right. Only then was a basic flaw in the design of the
MIC unit laid bare: it did not provide for even one safe
route for the gas to be neutralized at a very high
temperature and pressure before it escaped into the
atmosphere. Management compounded the original
design inadequacies by making a shortsighted equipment modification, less than a year before the accident, that may well have been responsible for allowing
water to enter MIC storage tank No. 610 where the
runaway reaction occurred. In addition, the few systems present that could have slowed or partially contained the reaction were all out of operation at the
time of the accident.[3]

A partial list of these defects follows:

- Gauges measuring temperature and pressure in
  the various parts of the unit, including the crucial

MIC storage tanks, were so notoriously unreliable that workers ignored early signs of trouble.

• The refrigeration unit for keeping MIC at low temperatures (and therefore less likely to undergo overheating and expansion should a contaminant enter the tank) had been shut off for some time.

• The gas scrubber, designed to neutralize any escaping MIC, had been shut off for maintenance. Even had it been operative, post-disaster inquiries revealed, the maximum pressure it could handle was only one-quarter that which was actually reached in the accident.

• The flare tower, designed to burn off MIC escaping from the scrubber, was also turned off, waiting for replacement of a corroded piece of pipe. The tower, however, was inadequately designed for its task, as it was capable of handling only a quarter of the volume of gas released.

• The water curtain, designed to neutralize any remaining gas, was too short to reach the top of the flare tower, from where the MIC was billowing.[4]

These failures the night of December 2, 1984, left the skeleton work crew on duty helpless in the face of the massive MIC leak that occurred. The leak was first detected by workers about 11:30 P.M. the way they usually discovered MIC leaks: their eyes began to tear and burn. (Union Carbide acknowledged after the MIC tragedy that workers in its West Virginia plant used the same method—watering eyes—to detect the MIC leaks, which also frequently occurred there, despite the computerized system.⁵)

Workers in the Bhopal plant later told reporters that when they informed their supervisor of the leak, which was initially noticed as a drip of liquid and some yellowish white gas collecting about 50 feet off the ground, he said that they would deal with it an hour later, after their tea break. By 12:40 A.M., when workers returned to the area of the leak, temperature and pressure had built up to such a degree inside the tank that a 60-foot concrete slab at least six inches thick next to the tank started vibrating and then cracked. A loud hissing sound signaled the escape of the gas out of the top of a tall smokestack and into the night air over Bhopal.⁶

Soon after, the first victims arrived at the local hospitals and clinics; then they came by the hundreds, thousands, and finally hundreds of thousands. By dawn, the dead lay everywhere in the streets of Bhopal, their bodies bloated. Nobody counted as the bodies were heaped in piles and cremated according to Hindu tradition, or wrapped in shrouds and buried according to Moslem tradition. Posters with numbered pictures of some of the dead were put up so that relatives and friends could try to identify them. Many remained unidentified, however, since entire families had vanished in the catastrophe, with no one left to identify or remember them. Therefore, only an estimate of the toll is left for historians—200,000 exposed by most approximations; at least 2,500 dead and 17,000 permanently disabled.[7]

For many of the living, however, the horror had just begun.

Notes

1. Sanjoy Hazarika, "Indian Journalist Offered Warning," *New York Times*, Dec. 11, 1984.

2. Ron Winslow, "Union Carbide Confirms That Problems With Tanks in India Were Found in '82," *Wall Street Journal*, Dec. 11, 1984.

3. "Bhopal: What Really Happened?" *Business India*, India, Feb. 25–Mar. 10, 1985.

4. Anil Agarwal, Juliet Merrifield, and Rajesh Tandon, *No Place to Run*, Highlander Center and Society for Participatory Research in Asia, 1985; see also Stuart Diamond, "The Bhopal Disaster: How It Happened," *New York Times* (a series starting Jan. 28, 1985).

5. Ron Winslow, "Union Carbide Moved to Bar Acci-

dent at U.S. Plant Before Bhopal Tragedy," *Wall Street Journal*, Jan. 28, 1985.

6. Diamond, "Bhopal Disaster."

7. B. Bowonder, Jeanne X. Kasperson, and Roger E. Kasperson, "Avoiding Future Bhopals," *Environment*, Sept. 1985, p. 10.

FIVE

—

The
Aftermath

June 1985, half a year later: In the shadow of the now-silent Union Carbide plant, 45-year-old Neelam Bai and her four children sleep under an open sky. They are too poor to afford even one of the ramshackle huts located nearby, and they survive on leftovers shared by their almost equally destitute neighbors.

Their only possessions of potential value in the world are three slips of green paper: the death certificates of Neelam Bai's 50-year-old husband, her mother-in-law, and sister-in-law. All three perished in the Bhopal tragedy six months before.

The green slips list Body No. 64, Body No. 96, and Body No. 245, all unidentified at the time of their "disposal" by rescue workers in the aftermath of

47

the Bhopal catastrophe. As the next of kin, Neelam Bai and her children are supposedly entitled to relief measures from the Indian government; but as of June, they have not received a single rupee in compensation.[1]

The plight of Neelam Bai and her children is shared by several thousand other victims of the Bhopal holocaust, as courts in the United States and India attempt to sort out the tangled legal web surrounding the world's worst industrial disaster, and as relief efforts are bogged down and bungled by government red tape and bureaucracy.

Many of the basic questions about the catastrophe have yet to be resolved, though experts in various fields continue to visit the site to investigate. The cause of the accident is generally believed to be the entry of water or some other contaminant into MIC storage tank No. 610, which in turn triggered an exothermic runaway reaction, causing the highly volatile gas to heat up, build up pressure, and eventually escape into the atmosphere.

But why was such a huge quantity of MIC stored at the Bhopal plant in the first place? Why were all the safety and back-up systems, inadequate as they were, shut down or malfunctioning on the night of the Bhopal disaster? Why was Union Carbide unable to provide accurate toxicity data on MIC or appropriate guidelines for medical treatment of the Bhopal victims? Why wasn't the Bhopal plant following the recommendations of the Union Carbide investigative team, which, two years earlier, had uncovered several flaws? Why didn't the insurance companies covering Union Carbide in the event of a chemical disaster require that the Bhopal plant fortify its safety systems? Why didn't the Indian authorities heed the repeated warnings sounded by Indian journalist Raj Kumar Keswani of the impending danger posed by the Bhopal plant?*

Adding fuel to the fire, Union Carbide officials, in March 1985, further complicated matters by suggesting that "sabotage" may have been involved[2]— the one circumstance that, if proved, might remove their legal liability for the accident. Furthermore, in the aftermath of the Bhopal tragedy, it became evident that the information vacuum about MIC extended beyond India and was in fact global. When interviewed in January 1985, officials at the various United Nations agencies in Geneva responsible for collecting data on hazardous chemicals admitted that they knew nothing about MIC before the catastrophe, and could come up with little data on MIC after the Bhopal incident.[3] Like the dead and injured at Bhopal, the whole world seemed to share an appalling state of

* The headlines on Keswani's articles would have been hard to miss: "Sage, Please Save This City"; "Bhopal on the Mouth of a Volcano"; and "If You Don't Understand, You Will Be Wiped Out."

ignorance about a toxic chemical now known as a mass killer of unprecedented proportion.

In this vacuum, citizens' groups in Bhopal started organizing the local residents to campaign for a future free of the kind of calamity that had shattered and permanently scarred their community. Tired of waiting for relief measures from the Indian government, poverty-stricken widow Neelam Bai joined one such group that campaigned for quick compensation for the Bhopal victims. In mid-June 1985, as one of a series of actions, 250 frustrated Bhopal victims seized land owned by Union Carbide and declared it a "free zone" for setting up a public health clinic.[4]

As local authorities started cracking down on protesters at Bhopal, jailing some and breaking up various meetings, halfway around the world in New York, Union Carbide was maneuvering to limit its liability for the disaster. By autumn 1985, over 130 lawsuits asking $100 billion in damages had been filed against Union Carbide in various U.S. jurisdictions; and another 2,700 claims, asking $3 billion in compensation, had been filed in India. (The Indian government passed a law stating that it would represent all of these claims.) The major U.S. suits related to personal injury and death claims were consolidated in a district court in New York by a judicial panel on multidistrict litigation.

But hundreds of individual personal injury suits, as well as other types of litigation, such as stockholder suits, were still pending—all stemming from the accident. The major question in all these cases was whether they would be tried in India or the United States—and a decision on that question finally came in May 1986 when a federal judge ruled that the consolidated cases (which by then numbered 145) would be heard in India.[5]

The government of India had filed suit against Union Carbide in the United States when settlement talks between the two parties broke down earlier in 1985, but the validity of that suit was weakened by evidence that showed that the Indian government shared a significant degree of culpability in the case.[6] By the end of 1986, the Indian government had obtained a court ruling that the company had to maintain at least $3 billion in assets to satisfy any eventual judgments against it.[7]

Meanwhile a string of near disasters at Union Carbide's U.S. plants (see chapter 12) in August 1985 left it a badly shaken company. Evidence of takeover attempts started to surface in the business press, 4,500 employees lost their jobs, and the company promised to commit $50 million to "clean up our act on emissions."[8]

About the same time, the A. H. Robins Company, makers of the notorious Dalkon Shield intra-uterine contraceptive device, which was responsible for numerous deaths and injuries around the world, filed for protection under U.S. bankruptcy laws. In doing so, the company was following the lead of Johns-Manville, the world's biggest asbestos manufacturer, which used a similar tactic to avoid its liability in the many cases of asbestos-linked diseases. Observers couldn't help wondering whether Union Carbide would eventually follow suit, saving itself and leaving most of the Bhopal victims uncompensated and forgotten. (See Appendix E, p. 197.)

The aftermath of the Bhopal tragedy kept what was the 37th-largest U.S. company in the public limelight far longer than otherwise would have been the case. Meanwhile, Union Carbide was trying to carry on its usual business activities. In September 1985, four walking, talking human advertisements could be seen tramping through the suburbs and outlying villages of Madras, the largest city in southeast India. The four men were wearing huge cardboard *Eveready* battery "gowns," promoting Union Carbide's best-known product in much the same way it has always been promoted in India. Inside their "gowns," the men were sweating, coated with dust. They reported that their pay for this work, which takes them 9 to 12 miles every day, rain or shine, is ten rupees (less than one U.S. dollar) per day.[9]

Notes

1. N. K. Singh, "Bhopal, Six Months After," *Express Magazine*, India, June 2, 1985.

2. Statement by Warren Anderson, Chairman, Union Carbide Corp. (Press conference in Danbury, Connecticut, Mar. 21, 1985.)

3. Personal interviews with the author, Geneva, Switzerland, Jan. 23–24, 1985.

4. "Bhopal Victims Occupy Union Carbide Land," *Japan Times*, Japan, June 5, 1985.

5. "Union Carbide OKs Conditions for India Trial," *San Francisco Chronicle*, June 12, 1986.

6. "Bhopal-Related Suits Merged in New York Court," *Chemical & Engineering News*, Feb. 11, 1985; see also "Bhopal, A Year Later: Union Carbide Takes a Tougher Line," *Business Week*, Nov. 25, 1985, p. 96.

7. "India and Union Carbide Reach Agreement on Bhopal Action," *San Francisco Chronicle*, Dec. 1, 1986, p. 15.

8. "Bhopal, A Year Later: Union Carbide Takes a Tougher Line," *Business Week*, Nov. 25, 1985, p. 96; see also "Union Carbide Cites Errors in Chemical Leak," *Wall Street Journal*, Aug. 26, 1985.

9. Sadanand Menon, "Eveready ads' exploitation," *Sunday Observer* (Bombay), Sept. 22, 1985.

PART
TWO

The
Syndrome

SIX

—

Nobody Knows

To *stop* that [chemical] race—it's a lost cause!
World Health Organization Official

Seveso, Love Canal, Three Mile Island—all sites of earlier industrial disasters—have left their names imprinted on the public mind as synonyms for environmental catastrophe. None of them, however, prepared us for the scale of devastation at Bhopal. The repercussions for world trade and industry, for national and international regulation, and for how people view the safety of their own communities will be felt for years.

But despite its far-reaching implications, the tragedy in Bhopal is still considered by many simply as an isolated event, unlikely to be repeated. The chemical industry points with pride to its safety record, which it says is better than that of industry as a whole.[1] But chemical experts tell us that a "risk-free" technological society is a fantasy. "As long as there are human beings on earth, especially designing, building, and operating equipment, the possibility [for disaster] always exists," says J. Charles Forman, executive director of the American Institute of Chemical Engineers. "There is no such thing as a 'zero-risk' society."[2] At the same time, however, industry leaders and analysts have been undertaking a serious re-evaluation of their manufacturing processes. There has been a certain amount of soul-searching by executives, who are now determined to regain a better public image by defusing popular and regulatory suspicion.

It is perhaps not surprising that no one better exemplifies this introspective process than Warren Anderson, former chairman of Union Carbide Corporation. Anderson flew to India right after the accident, was arrested and briefly held by authorities in Bhopal, and then turned the running of the company over to others while he dealt with the aftermath of the tragedy.

In a profile of the Union Carbide chairman written by *New York Times* reporter Stuart Diamond, Anderson said, "You wake up in the morning thinking, can it possibly have occurred? And then you know it has and you know it's something you're going to have to struggle with for a long time." Anderson and his wife, who have no children, stopped eating in restaurants after the tragedy, because, Anderson said, "I kind of felt that if somebody caught me laughing over in the corner over something, they might not think it was appropriate."[3]

Anderson admitted that he was caught between conflicting pressures while trying to maintain ultimate loyalty to his company. "If you listen to your lawyers you would lock yourself up in a room someplace. If you listen to the public relations people they would have you answer everything. I would be on every TV program." Company directors, shareholders, insurers, and government officials all had demands. At first, in the days following the accident, Anderson insisted that there were no differences between the Bhopal plant and Union Carbide's West Virginia plant; but in March 1985 he finally admitted that the doomed plant had violated company standards and operated in a way that would not have been tolerated in the United States.[4]

This admission was a tacit acknowledgment that there is indeed a "double standard" between the First and Third World operations of the multinationals, at least in the case of Bhopal. To Anderson and Union Carbide, however, the admission served two further purposes. One was to bolster their legal strategy, which would apportion the major share of blame to the Indian government for its failure to effectively regulate the plant and for allowing people to live nearby.

The second purpose was to reinforce, albeit in round-about fashion, the claim by Union Carbide and the rest of the chemical industry that pesticide manufacturing technology can be considered basically safe, as long as proper standards are maintained.

How inherently safe—or necessary—agrochemical technology may or may not be is a subject worthy of extended, informed public debate. Of the thousands of industrial plants now in operation, few have been subjected to rigorous public scrutiny. During 1984, for example, DuPont announced plans to build new pesticide plants in Indonesia and Thailand; Hoechst, in India, Pakistan, and Colombia; both Stauffer and Sandoz, in Brazil; and Monsanto, in Taiwan.[5] Each year brings new announcements of similar plans by companies, big and small, throughout the world. Rarely is the public involved in reviewing these plans for the potential hazards they present to their communities. And often the authorities in host countries seem uninformed about the technologies being introduced into their environment and the safeguards necessary to keep them under control. India's late Prime Minister Indira Gandhi, for instance, once stated, "Environmental safeguards are irrelevant: poverty is our greatest environmental hazard."[6] In this context, it is essential that officials in the international agencies keep track of the proliferation of chemical plants, catalog the manufacturing processes involved, and evaluate the possible health effects of the compounds in common use.

One of the major centers for this work is Geneva, Switzerland, headquarters for some of the extensive United Nations bureaucracy that tries to help safeguard the world from the dangers of pesticides and other chemicals. One current project is the International Program on Chemical Safety (IPS), a joint venture of three U.N. offices: the International Labor Organization (ILO), the World Health Organization (WHO), and the United Nations Environment Program (UNEP).

In the days following the Bhopal tragedy, Georg Kliesch, ILO's official with IPS, searched for references to MIC in his agency's Encyclopedia on Safety and Health, which has over the past 50 years chronicled chemical dangers and has more than 2,000 extensive articles. Kliesch was shocked: "You don't even find MIC in there!" Upon further searching, Kliesch discovered that this information gap extends for virtually all the intermediates used in pesticide production. "It [the Carbide MIC process] is not a major process in the industry worldwide but rather only one of many dangerous problems. Bayer, for instance, has its own process with MIC. They do not store it at all, but simply produce it as they need it in the production of carbamate pesticides." This so-called continuous process avoids the dangers of storing MIC so tragically realized at Bhopal, but according to Kliesch, "nobody can necessarily say the Bayer process is safe. At this point in the ILO, if there are so many different chemicals in so many different chemical processes . . . well, you just try your best."[7]

Nearby, at WHO, Gaston Vettorazzi is equally pessimistic: "We know nothing about the health effects of intermediates and very little about the finished pesticides themselves. But I am afraid we will never be able to slow down this race to use chemicals in the world enough to find out what we need to know. And to *stop* that race—it's a lost cause!"[8]

Another U.N. official, who did not want to be identified, said that his investigations of multinational investment in Third World countries reveal a pattern of "shortcuts" by the companies that leads to potentially dangerous situations. "Even those companies that say they will maintain the same standards as in the developed world find it difficult to resist the temptation to take a shortcut. Even if they have a good design for their plant, however, there's no good infrastructure in the underdeveloped countries. Even if they put it away from population centers, who will check and control that the people don't come in around it? Who will check the standards of construction? Most underdeveloped countries do not have the expertise to do these things.

"The workers must have good training; you cannot play with these things. You must be very careful—somebody must be checking the whole day long and the government must check at irregular [*sic*] intervals. This is where a lot of things go wrong, especially in the Third World."[9]

Jan Huismans, the UNEP official responsible for maintaining the International Registry of Potentially Toxic Chemicals (IRPTC), says that when the accident at Bhopal occurred, "MIC was not on our list. In fact none of the intermediary chemicals were on it because the governments themselves don't know them. We are able to add them whenever a government identifies them as of concern." Huismans says he has seen hazardous pesticide plants all over the world, "sometimes right in the middle of a town. In Africa, for example, they start with a little planning and try to locate these plants outside a populated area. But in no time, these cities grow and the industrial areas are engulfed by population settlement, surrounded by shantytowns.

"Also, there are no adequate waste disposal facilities for these plants. There is a lack of awareness generally about how dangerous pesticides are. There's a lack of skilled regulatory personnel and controls. There is, in sum, a whole syndrome of problems."[10]

This, then, is the Bhopal Syndrome. It is time to take a firsthand look at some plants that have it.

Notes

1. Cathy Trost and Carol Hymowitz, "Congressmen, Environmentalists Fear that Laws in U.S. Wouldn't Prevent a Poison-Gas Disaster," *Wall Street Journal*, Dec. 6, 1984.

2. Personal correspondence with Daphne Wysham, Center for Investigative Reporting, July 24, 1985.

3. Stuart Diamond, "Warren Anderson: A Public Crisis, A Personal Ordeal," *New York Times*, May 19, 1985.

4. Ibid.

5. Wood, Mackenzie & Co. Agrochemical Service, 1984.

6. Claude Alvares, "The Bhopal Tragedy: Climax of a Hazardous System," *Asian-Pacific Environment*, Penang, Malaysia, March 1985.

7. Personal interview with the author, Geneva, Switzerland, Jan. 24, 1985.

8. Personal interview with the author, Geneva, Switzerland, Jan. 24, 1985.

9. Confidential interview with the author.

10. Personal interview with the author, Geneva, Switzerland, Jan. 23, 1985.

SEVEN

———

Slow-Motion Bhopal

The roofs in the village and the yards are coated with
DDT dust.   ICI Official

Cicadas, Indonesia: a picturesque village area in the
tropical hill country south of Jakarta on Indonesia's
main island of Java. A small sign shaped like an ar-
row, with the name P. T. Montrose P.N. stenciled on
it, points the way down a dirt road to a clearing where
two pesticide formulating plants stand, surrounded
by fences with gates manned by guards.[1]

Since late summer, 1984, the villagers at Cicadas
say, they have been suffering from a sort of slow-
motion Bhopal. The plant operated by P.T. Mantrose* Pestindo Nusantara (which is affiliated with a

* Montrose Chemical Company of California dismantled its DDT
plant in Torrance and in 1983 shipped much of the manufacturing
equipment to Indonesian-owned P. T. Mantrose. The two firms are
not otherwise affiliated.[17]

U.S.-based firm, Barsky Sales), supplies a significant portion of the world market for water dispersible DDT powder.[2]

Starting in mid-August 1984, the Cicadas plant began burning its waste products at night from time to time, producing smoke and dust, which blanketed the area. Soon after the first burning, the villagers say they started to exhibit symptoms of chemical poisoning, including the marked shortness of breath, similar to those that would be experienced by the Bhopal victims several months later. Some of their domestic animals, primarily goats and cows, were said to have died after grazing on contaminated grass. Then, reportedly, the villagers themselves began to die: choking, vomiting, and hypersalivating in the familiar litany of poisoning symptoms caused by pesticides. The numbers of dead steadily mounted until, by June 1985, 25 villagers were reported to have died. At that point, investigators from two citizens' groups in Indonesia—WALHI (Indonesian Environmental Forum) and KRAPP (Indonesian Network Against Misuse of Pesticides)—uncovered the DDT poisonings at Cicadas and demanded that the authorities take measures to remedy the situation.

According to Gleb Barsky, president of Barsky Sales in Sandy Hook, Connecticut, "the government formed a high-caliber commission that came in and stayed at our [Cicadas] plant site for five to seven days and reviewed the entire operation and made recommendations. They said, 'Get a better incinerator,' which I had been saying for years, and then they gave the plant a clean bill of health."

Barsky, who calls himself "the world's last American DDT salesman," offers a different explanation for the controversy at Cicadas: "The report of 25 deaths is an exaggerated claim by the local village headman because our plant manager refused to pay him off. In Indonesia, if you're a plant manager, you have to pay the village headman every month or you'll have problems. The headman at Cicadas wanted from our plant manager a couple hundred dollars a month or whatever. But the plant manager is a stubborn SOB and he said no. All of a sudden, chickens started 'dying' and then every death, sickness, or miscarriage in the village was our fault.

"[But] I don't see the connection. It was only waste bags that had contained DDT that were being burned. There is not enough DDT in those waste bags to get a mosquito to sneeze.

"Now, if you burn DDT, you can release phosgene unless it's done right," Barsky continues. "But as long as the temperature is high enough, there's no problem. Was there visible smoke? Yes. Was there a bad odor? Yes, no question about that. I insisted that they clean up their act, get a better incinerator, and a higher chimney."

Barsky added that "P.T. Mantrose is fully Indonesian owned. . . . One of the partners is the son of

President [Suharto] of the republic. The chief financier and founding light, really, of the company is Chinese. But the plant is located in a Moslem fundamentalist area. The Moslem fundamentalists are giving Suharto a hot foot to bring in Moslem law as opposed to open religion, which the president supports. The fundamentalists cannot attack him directly because he's fairly powerful and a ruthless dictator, really, so they try to get at him through his sons at this plant."

Barsky says the DDT formulated at the plant is used exclusively in malaria-control programs in Indonesia and other Third World countries. "If you look at the incidence of malaria in the world, you'll see why we need [DDT]," Barsky says. "[But] the countries that have malaria are the poorest countries in the world. They're corruptly ruled and would rather spend their money on 747s and guns than on malaria control.

"Bhopal has two sides to it," Barsky continues. "The whole insecticide question has two sides to it. There are manufacturing processes overseas in Third World countries that are detestable. Many places I visit, I hate to go in the factory. . . . The host governments share responsibility. . . . I'm not exonerating the companies from all responsibility. They're perfectly happy to pull out their profits, but the governments share the blame, too."

Barsky says "the DDT sales business used to be a

clean, aboveboard business. [But] we are competing now for a 75 metric ton order in Africa. The U.S. AID [Agency for International Development] is providing the financing. In the past, AID would have bought DDT through the GSA [General Services Administration], and everything was in the open. But now they are not working through the GSA. Instead, they have a pan-African front organization to handle the purchase of DDT because they don't want to be associated with it. That means there is a double overhead and it's wasteful. They are using funny African purchasing agents. It's a dirty, shitty business."[3]

The pesticide formulation plant right next door to the Mantrose DDT plant at Cicadas is owned jointly by the U.K.-based ICI, the world's fifth biggest pesticide producer, and Agrico, an Indonesian national company. The plant employs 60 workers and annually formulates 5,000 tons of powder, liquid, and granular solutions of insecticides, herbicides, and rodenticides. Its best-known product is the herbicide paraquat. ICI headquarters are on the tenth floor of an office building in a busy section of Jakarta. Outside, poor people crowd the gates. The elevator up to the ICI offices passes the fourth floor, marked not by a number but by the famous yellow scallop shell logo of another pesticide giant, Shell.

"Yes, there are slums all around our plant," acknowledges ICI Work Manager M. A. Nasution,[4] sadly. "The people live even between the fences of our plant and the one operated by Mantrose. When we first built this plant two years ago, nobody was around there then. But after a year and a half, lots of people were around there. [The village headman denied this to WALHI and KRAPP, saying the villagers lived there before ICI and Mantrose opened.] Some of the poor sit in front of the factory as vendors, hoping to sell little things to the workers when they come and go."

Since there is some danger of water pollution from the plant, ICI started monitoring the village wells. "But it makes them suspicious," says Nasution. "So now we pay the Department of Health a little extra every month and they do the sampling for us. Pollution control is always a problem in agrochemical plants because every five or ten years a new product has to be developed [due to various factors, including changing market conditions, pest resistance, and regulatory actions]." "People just don't know about pesticides," Nasution says. "They don't understand. Agrochemicals are difficult to defend."

M. S. Sitanggang, ICI's marketing manager, explains that the weed killer paraquat "mostly goes to large estates . . . growing rubber, oil palm, tea, and cocoa."[5] It is on these plantations, and elsewhere in southeast Asia, that paraquat has become a major cause of death, both accidental and suicidal.[6] Some of the accidental cases, Sitanggang feels, could be avoided: "This is my own idea but I think people tend not to read all the information on the label. Probably symbols would be better. Labels are probably not an effective way to protect people from poisoning."

ICI had another poisoning problem recently that underlined the tragic disparity between the high-technology world of the multinationals and the desperate reality in the fields of Indonesia. The company's rat killer, *Klerat*, a lump of rice coated with poison, was intended for use in field and storage areas where the rodents cause damage. During a drought in 1983, however, starving villagers in several parts of the country gathered *Klerat* and tried using it for food. Knowing it was a poison, they washed it repeatedly before cooking and eating it. Nevertheless, in Lampung, Sumatra, 20 people died; in Musi Banju, three more succumbed; in Kemcamatan, West Kalimantan, two others.[7]

Now the company is trying a new method of coating the rodenticide in wax, but at least one corporate official has doubts about the use of *Klerat* at all. "[The government] approved ICI's new, wax-covered rice with *Klerat*, but there's a possible problem," explains Anwar Arif, of the U.S.-based multinational Monsanto. "The rats eat it, die, rot in the field, and attract flies, which spread disease to humans. And perhaps the snakes and birds will also be affected."[8]

Problems such as suicides, unlabeled containers, illiterate workers, chronic misuse, and hungry people eating rat poison raise the question of whether pesticides are an appropriate technology for Third World societies like Indonesia. Nevertheless, the Indonesian government heavily subsidizes the use of agrochemicals, by serving as an intermediary between producers and farmers, and absorbing much of the cost. In 1983, for example, the government bought diazinon from the producer at a price of 6,169.50 rupees per liter (rpl) and sold it to farmers for 1,230 rpl.[9] By fiscal year 1986–7, pesticide and fertilizer subsidies cost the government 671.5 billion rupees, though for 1987–8, the amount was cut to 203.5 billion rupees.[10]

"In the developing countries, and this is my own opinion, we don't have such good safety consciousness," says ICI's M. S. Sitanggang. "Just look at the streets out there." Outside his office, every manner of vehicle—from buses and trucks belching black smoke to autos and bicycles and mopeds and rickshaws and carts pulled by animals or humans—weaves in and out of each other's way, often with sudden—and bloody—consequences. The route from ICI's office to its plant runs through the densely populated Cibubur district on the edge of Jakarta. Here, among what seem to be wall-to-wall shantytowns filled with poor people—squeezing through dirty alleys that pass for streets and along filthy ditches that serve as water supplies—rise the shiny factories of multinationals like

Ciba-Geigy, Fairchild, Warner Lambert, and Union Carbide. In the shadows of these steel and concrete structures people live in hovels as insubstantial as cardboard boxes.

It was inside Union Carbide's battery plant on a rainy night in April 1978, that a worker named Haryanto died. As revealed by reporter Bob Wyrick in *Newsday*, Haryanto's feet were wet from water spilled on the factory floor from a leaky drain. His body was black with carbon dust from the "mixing room" where he worked; the machine he ran mixed carbon black with other chemicals used in *Eveready* batteries, but the room's two dust collectors were broken and had not been repaired. He had been required to work three consecutive overtime days and was "too tired to think properly" at the time of his accident, according to the examining physician. When Haryanto tried to depress an improperly located and unshielded electrical switch with a metal saw blade, he was electrocuted. Conditions overall at the plant were so bad that the company doctor reported that 402 workers, over half of the work force of 750, were suffering kidney disease from occupational exposures to dangerous substances.[11]

"In the developing countries, ignorance is really the key," argues Monsanto's Arif.[12] "[The government] produces good on paper, but supervision, enforcement is a problem. There is no money to do it.

There is not even any money to train the members of the Commission about what [they] are supposed to be doing."

Arif feels that a governmental double standard exists also, which sets a bad example. "We have discovered that DDT has illegally found its way into mosquito coils, which means that somebody in the government acted improperly. How can we expect nongovernment people to follow our regulations when the government people themselves do not follow them? The government gives itself an [emergency exemption] whenever it wants to use a certain pesticide that has been banned."

Arif is also concerned about the danger from the pesticide plants. "After all, Bhopal was partly the government of India's fault because they allowed all that squattering to occur around the plant. It's happening here also."[13]

Despite its many pesticide-related problems, Indonesia may escape a disaster of the immediate scope of Bhopal, simply because its market for pesticides, though growing fast, is not yet as big as in countries like Brazil or India. "The Indonesian market size does not currently justify a manufacturing plant [as opposed to the formulation facilities now operating] from the multinational point of view," explains ICI's Sitanggang. "That reality minimizes the risk of a 'Bhopal' happening here. It's a kind of natural barrier."[14]

Nevertheless, only after the Bhopal tragedy did it become clear how close Indonesia might have come to a disaster. In late December 1984, the government suddenly postponed approval for two carbamate pesticide plants that would have used a Bhopal-type, MIC-based process in Indonesia.[15] And internal Union Carbide documents made public during court proceedings in New York reportedly revealed that, shortly before the Bhopal disaster, the company had considered dismantling its MIC plant and moving it to either Brazil or Indonesia.[16]

Notes

1. Personal observation by the author, Cicadas, Indonesia, June 10, 1985.

2. National Agricultural Chemicals Association, Washington, D.C.

3. Telephone interview with the author, Dec. 10, 1985.

4. Personal interview with the author, Jakarta, Indonesia, June 10, 1985.

5. Personal interview with the author, Jakarta, Indonesia, June 10, 1985.

6. R. D. Amarasingam and A. S. Lim, "Review of cases of human poisonings recorded from 1977–1981," Toxicology Division, Department of Chemistry, Malaysia; see also: K. W. Chan and K. S. Cheong Izham, "Paraquat

poisoning: A Clinical and Epidemiological Review of 30 Cases," *Medical Journal of Malaysia*, vol. 37, no. 3, Sept. 1982.

7. "Pesticide Poisoning Incidences in Indonesia Between 1983 and August 1984," Directorate of Hygiene and Sanitation, Jakarta, Indonesia.

8. Personal interview with the author, Jakarta, Indonesia, June 11, 1985.

9. "Prices of Pesticides Procured from P. T. Petrokimia Kayaku 1982/3," Agriculture Ministry, Jakarta, Indonesia.

10. Vandine England, "Bugs in the System," *Far Eastern Econ. Rev.*, March 19, 1987, p. 117.

11. Bob Wyrick, "How Job Conditions Led to a Worker's Death," *Newsday*, Dec. 17, 1981.

12. Personal interview with the author, Jakarta, Indonesia, June 11, 1985.

13. Personal interview with the author, Jakarta, Indonesia, June 10, 1985.

14. "Proses Produksi Pabrik Pestisida Milik Union Carbide Harus Diubah," *Stnar Harapan Daily*, Jakarta, Dec. 29, 1984; see also: "Izin 2 industri pestisida dicabut BKMP," *Pos Kota Daily*, Feb. 6, 1985.

15. "Carbide 'Was Poised to Dismantle Bhopal Unit,'" *European Chemical News*, Oct. 7, 1985, p. 21.

16. John Kallok, former plant manager, Montrose Chemical Co. of California, telephone interview with Andrew Porterfield, Center for Investigative Reporting, May 27, 1987.

EIGHT

Run into the Wind

> If a [phosgene] leak occurred today, people would die.
> It would enter their houses; they would notice a sweet-
> sour smell, but bearable . . . go to sleep and die.
>
> A former Bayer supervisor

The people living down wind and down river from the massive Bayer industrial complex in the Belford Roxo district, some 20 miles from downtown Rio de Janeiro, know a lot about chemicals. Inside the mammoth German-owned industrial site are more than a dozen separate factories, manufacturing everything from acids and chrome salts to dyes and resins, from pharmaceuticals and emulsifiers to phosgene and isocyanates closely related to MIC, plus a long list of pesticides, including ethyl and methyl parathion, the nerve gas derivatives that originated in the German laboratories of World War II.

Approximately 1,800 people work in the complex, many living among the tens of thousands who exist "mainly in poor and even shantytown neighborhoods," according to David Hathaway, author of an upcoming book on the pesticide industry in Brazil. The workers from the dye plants come home a different color each day, "sometimes yellowish, other days blue or red—depending on the pigment in use."[1]

The wives and mothers of workers in the plants get used to the products their husbands and sons work with. "There is a lot of pollution and a bad smell, a chemical smell that comes at certain times of the day," explains Maria das Gracas, who lives near the Sarapui River flowing beside the plant. "I know the smell because my husband works in the laboratory with chemicals for dyeing clothes and I am familiar with all kinds of chemical smells." Das Gracas spoke with documentary producer Frederico Fullgraf in June 1985, about the pollution coming from the Bayer plant. "There are certain hours when they let out that smell, about ten in the morning and again around six or seven in the evening. . . . It is an acid smell that burns. This very morning I woke up with my eyes irritated; they are not red, but they burn. And my daughter's eyes are actually red."[2] Das Gracas says the river by the factory "turns red, real red. . . . Sometimes it stays red for three, four days. And it stinks!" Another resident, Rafael Luis says, "It looks like they threw blood into the river."[3]

"Ever since [the Bayer complex] came, we have

been having these problems," says das Gracas. "This river had fish in it, but not any more, none at all. My mother used to fish in this river, and it's been years now that the fish are gone." Besides pollution from Bayer and other chemical factories in the area, residents throw trash and garbage in the river, compounding the problem into a serious public health menace. "They say the garbage truck comes by, but it never does," explains das Gracas. "And we have no storm sewers on the street. This whole riverbank area floods, the water flows from up there and right into our houses, full of pollution from the river. And when people step in it their feet get full of sores, and that water is rotten."

The area, which was once quite pleasant (residents used to swim in the river during the 1950s), now seems barely habitable to das Gracas. "When the sun comes out," she says, "you smell the chemicals mixed with the smell of rot coming out of the river mud. I lock myself into the house, because I can't stand being outside with that horrible smell."

In the midst of this poor, underdeveloped community, Bayer's modern chemical works churns out a large share of the company's $265 million (in 1983) worth of Brazilian products, which, though impressive, represents less than 2 percent of the multinational's global sales.[4] Bayer is one of the most internationalized companies in the world, garnering 79 percent of its total sales from overseas business.[5] Pesticides account for a little over 11 percent of the group's total sales and hit the $1.5 billion mark in 1983, meaning that Bayer alone controls almost 10 percent of the world pesticide market.[6] In Brazil, pesticides are produced for cotton, tomatoes, potatoes, coffee, soy, wheat, green vegetables, citrus and other fruit, tobacco, corn, cocoa, rice, cucurbitaceae, rubber, and many other crops.[7]

Sebastian Richard, Bayer's manager for environmental affairs at Belford Roxo, rejected residents' claims that the company's complex is responsible for pollution in the area. Interviewed by David Hathaway in December 1985, Richard stated that Bayer opened an $8 million biological treatment plant in October 1984 to control all solid and liquid effluent from the plant.[8] According to Richard, only treated liquid wastes flow into the river now, and all solids are deposited in a 60-acre landfill lined with clay and polyurethane. As to the red color of the river, "that I can say is technically impossible," Richard stated. "There is only one single outlet for all our [treated] wastes. So there is no technical possibility of any given plant in the [complex] bypassing the treatment station."

Former Bayer supervisor and trade union leader Paulo Morani gave a different sort of insider's account of working conditions in the Bayer complex.[9] Interviewed by Fullgraf in June 1985, Morani said, "the pesticides plant used to smell terrible and the workers at that plant had to shower and change clothes before entering the lunch hall. They only wear masks in the plant when they are going to handle vaporous chemicals that are visibly dangerous." Morani says a visitor to the plant would see "pesticides leaking down the middle of the road. People come and clean it up, but it is handled like that. They are highly corrosive and eat through the tubing if it is not the right kind."

Morani, who until mid-1983 was production foreman for the polyurethane plant in the Bayer complex, explains some of the hazards he witnessed: "In my area [we had] the chemical reaction in four reactors working with propylene oxide [a chemical that decomposes at temperatures above 71.6 degrees F, provoking an irreversible chain reaction that leads to an explosion]. This chemical is in tanks, covered with cold water. One time there was no cold water for 24 hours due to a problem in the plant's water cooler and a barrel exploded when being handled by safety personnel, sending the Germans into panic."

There were further reminders of Bhopal in Paulo Morani's unit. "We also handled toluene diisocyanate [TDI], a cousin of MIC, and also highly toxic." Morani says the chemical is shipped from another complex in Brazil.

According to Morani, and also to a March 1985

report issued by the Rio de Janeiro State Environmental Studies Foundation (FEEMA), the main danger to the surrounding community may be the unit's manufacturing of yet another member of the isocyanate family, methyl diphenol diisocyanate (MDI), which requires two extremely hazardous intermediate products—phosgene and chlorine. Both are highly toxic, and there have been fatal chlorine leaks in many parts of the world. Chlorine and one of phosgene's components, carbon monoxide, are stored at the plant; either gas, if leaked, could kill by asphyxiation. Morani says the carbon monoxide is stored in a spherical tank next to the railway tracks.

The main fear, however, is phosgene. "There are three levels of security alert in the MDI plant," Morani explains. "At the first level the MDI plant would be evacuated. . . .

"At emergency level two, as it was known at the time I learned the procedures, we would have to evacuate the entire factory complex, in the case that so much phosgene had leaked that all plants had to be cleared.

"And at level three, we would have to evacuate the population of Belford Roxo—hundreds of thousands of people. And, depending on the winds, the phosgene can travel up to six miles, threatening up to a million people, in my estimate.

"Nobody is aware of this danger," Morani continues, "probably not even the majority of the workers in the MDI plant itself. . . . I believe that if a leak were to occur today, a major one, people would really die. And worse yet, [whereas] at Bhopal the chemical could be detected, it burned. . . . But not phosgene. It would enter people's houses; they would notice a sweet-sour smell, but bearable—so much so that the Jews didn't scream out as they were being killed with it in the Nazi gas chambers; they went to sleep and died. And that is what will happen [here]."

Residents say they have never been told by Bayer of this danger. All they know, they say, is that in an emergency situation they are to look to the 20 or so wind socks hung around the plant and run in the opposite direction, i.e., into the wind.[10]

Asked to comment on Morani's charges, Bayer's Sebastian Richard stated, "Half of his affirmations are fantasy. We have plants with the most modern know-how in construction, process, and safety." More specifically, in an interview, the Bayer official—

- rejected as "hearsay" the charge that workers in the pesticides unit showered and changed clothes before lunch;

- stated that pesticide spills on the street are "impossible" because of Bayer's precautions and compliance with regulations;

- was unaware of any propylene oxide explosion;

- rejected the danger of MDI, phosgene, and chlorine leaks ("In the MDI plant, we use small quantities of phosgene. It is used immediately and there's no storage. The operators know what to do in the remote case of a leak. . . . We only have small amounts of chlorine gas stored, only what is strictly necessary to produce phosgene.");

- stated flatly that "we have absolute certainty that our safety systems in all plants are closed [leakproof]. It is absolutely impossible for any gas produced in those units to reach any place outside the factory";

- stated that the "levels of emergency do not exist." Said the wind socks around the plant "have no emergency function but an operational utility in plants whose operating temperature can be affected by changes in the wind." Said Bayer has no emergency evacuation system, because "Belford Roxo has 400,000 inhabitants. It would be insane for Bayer to have an alarm system for evacuating a population of 400,000."[11]

Therefore, however remote the possibility of an emergency at the complex, residents would indeed be left to look to the socks and run into the wind.

Though Bayer is the biggest pesticide company in Brazil—controlling 14 percent of the market in

1982—most of the other major world producers are there as well. Currently, about 50 companies have manufacturing or formulating plants seeking to tap what has become the world's fourth largest pesticide market.[12] After the Bhopal accident, public attention was focused on the dangers represented by MIC. State officials turned back a shipment of 13 tons of MIC from Union Carbide (U.S.) to Elanco Quimica, a pesticide manufacturing company in Rio de Janeiro owned by the U.S.-based multinational Eli Lilly.[13] In addition, Union Carbide itself had a carbamate plant in Brazil using MIC, and the U.S.-based FMC was scheduled to begin using the substance until it was banned by local authorities in the wake of the Bhopal tragedy.

If there is going to be "another Bhopal," it seems likely that it could happen in Brazil, with its sprawling industrial complexes and fast-growing population. Accidents are, after all, bound to happen. On February 25, 1984, for example, in Cubatão (known as "pollution valley"), near São Paulo, gasoline leaking from a pipeline exploded, setting off fires in a nearby shantytown. By the time the fires were put out, at least 500 people were dead.[14]

Notes

1. Personal interview with Frederico Fullgraf, Nova Iguacu, Brazil, June 1985.

2. Personal interview with Frederico Fullgraf, Nova Iguacu, Brazil, June 1985.

3. Personal interview with Frederico Fullgraf, Nova Iguacu, Brazil, June 1985.

4. Associacao Nacional de Defensivas Agricoles, Brazil.

5. Jess Lukomski, "Bayer Profits Jump On Gains Abroad," *Journal of Commerce*, May 8, 1985.

6. Wood, Mackenzie & Co. Agrochemical Service, 1984.

7. Associacao Nacional de Defensivas Agricoles, Brazil.

8. Personal interview with David Hathaway, Belford Roxo, Brazil, Dec. 19, 1985.

9. Personal interview with Frederico Fullgraf, Nova Iguacu, Brazil, June 1985.

10. Personal communication from David Hathaway to the author, July 31, 1985.

11. Personal interview with David Hathaway, Belford Roxo, Brazil, Dec. 19, 1985.

12. Wood, Mackenzie & Co., 1984.

13. Associated Press, "Brazil Returning Poison Gas," *San Francisco Chronicle*, Dec. 11, 1984.

14. "List of Major Industrial Accidents Over the Past Seven Decades," *Consumer Interpol Focus*, Penang, Malaysia, June 1985.

NINE

The Neighbors Strike Back

We built in 1968 and no people were here. Now the people come closer and closer and we are bothered by this. The waste air made some people sick.

Entomologist, Cheng Hong Chemical, Ltd.

Taiwan is a beautiful tropical island with terraced rice fields climbing the sides of its interior mountains and a population of 19 million jammed into its urban areas. There simply is not enough space for all the infrastructure necessary to meet the population's basic needs, including the production of food. Thus, the island's 66 small pesticide producers, most of whom are local formulators using technology licensed from the multinationals, operate right next door to people's houses, competing for the limited supplies of space, water, and clean air.

Under these circumstances, conflict is inevitable. Taichung is a good case in point. A bustling city of

600,000 people in the middle of Taiwan, Taichung is a market center for the surrounding agricultural districts—peasant women wearing traditional bamboo hats push fruit and vegetable carts into the city in the early morning—and it is where the country's major agricultural research station is situated.

Taichung is also the site of a number of Taiwan's pesticide plants, like Sunko Chemical's paraquat formulating factory. Neighbors of the Sunko plant complained of the air pollutants emitted, which made them feel dizzy and nauseated. But protests to the plant management and government officials failed to bring relief, they claim; so they decided to take matters into their own hands. On the night of May 13, 1985, a group of more than 100 angry residents, calling themselves an "antipollution team," stormed the plant, breaking most of its windows and shouting epithets, thereby attracting the attention of the nation's press.

In response to public and media pressure, the government's Provincial Environmental Protection Bureau (PEPB) tested the air around the plant and found evidence of a pyridine compound, which is paraquat's key intermediate, and found other contaminants in the water in excess of water quality standards. Based on these findings, PEPB ordered Sunko Chemical to stop production.[1]

In December 1984, in the aftermath of the Bhopal

tragedy, PEPB inspected all the country's pesticide plants and found five that were seriously hazardous. One was Cheng Hong Chemical, located on the outskirts of Taichung. The Cheng Hong plant, identifiable by its strong smell from some distance away, sits at the end of a narrow road surrounded by coconut palms and banana trees. Outside, small, uniformed children skipped past on their way home from a nearby primary school.

Teh-Chi Chung, an entomologist in the company's development department, explains how the plant got into trouble. "We used to synthesize benlate and paraquat, but we stopped [under government order] because of air and water pollution and because two of our workers got skin cancer. Paraquat's intermediate pyridine caused the cancer, a doctor at the Veteran's Hospital discovered. He say that three to five years exposure to pyridine causes the skin cancer. But our workers are all right now except for the scars."

Chung said he worries because "people are living too close to the plant. . . . We built in 1968 and no people were here. Now the people come closer and closer and we are bothered by this. The waste air made some people sick. . . ."[2]

Across town sits Taiwan's largest pesticide producer, Shinung Corporation, which has been formulating pesticides for 30 years and manufacturing them for the past 15 years. Most of its output is herbicides, but it recently added fungicides and insecticides to its product inventory. Ching Ling Chiou, section chief of the company's pesticide department, explained how Shinung guards against the possibility of a Bhopal-type accident in its plant—one of three using MIC in Taiwan.

"We import technical products from Bayer and use a much different process from UCIL [Union Carbide India Ltd.]. We combine dimethyl sulfate and sodium cyanate [a solid] to create MIC. In India, of course, they combined phosgene and methylamine gas under high pressure. Our process is more expensive than the phosgene process."[3]

Shinung uses MIC to make a series of carbamate pesticides, including carbofuran. "Our MIC production capacity is one ton a day, but we don't store it, even in the manufacturing process," Chiou explains. In case of a leak, "we have an automatic alarm and a water curtain that will neutralize any MIC. And our emergency power supply will start even if there is a power failure. Our waste air and waste water are checked by the [Taiwan] EPA. We have a gas scrubber, washing tower, and active carbon absorption.

"In Taiwan, the production of phosgene is prohibited," Chiou continues. "Phosgene is very useful and cheap in the production of a number of products besides pesticides—polyurethane, for example. In deciding which intermediates are safe, toxicity is one factor, but there are others as well, such as the residual rate. And if the boiling point is very low, like with MIC, it is very dangerous to store and use.

"At the MIC plant we have three alarm systems—one for the residents, one at our office, and one around the plant as a whole. We have an information poster for the residents. In a first degree alarm, we can control it ourselves. In a second degree alarm, we cannot control it; so the public alarm is sounded and we inform the fire and police departments. The area would be sealed off."

The plant has 60 workers in the formulation section and 20 in the manufacturing section. During a tour of the facility, the only worker wearing protective gloves in the section packaging pesticides is a young woman who, Chiou explains, "wants her hands to stay nice." S. C. Roan, assistant manager in Shinung's agrochemical department, explains that the workers are checked for evidence of toxic exposures once a month. They also drink a kind of vitamin treatment and milk every day, according to Roan, to compensate for chronic toxic exposure, though there is scant medical evidence that this works. Roan himself, according to recent medical tests, has a "slight poisoning problem."

Roan says "it's not the company's fault people have moved so close to the plant. The government doesn't enforce the law that they should stay a certain distance away."[4] Shinung's corporate group, which originated with pesticides but now includes a number of other businesses, operates a food plant, just over 200 yards from the pesticide plant, which prepares box lunches for factories, schools, and other institutions.

One of the people most knowledgeable about pesticide manufacture and use in Taiwan is Dr. Gwo-Chen Li, director of the Taiwan Agricultural Chemicals and Toxic Substances Research Institute, and president of the Plant Protection Society of the Republic of China. Dr. Li explains why people are living so close to Taiwan's 66 pesticide factories: "The law says there must be a 40-yard distance between the edge of the factory itself and the borderline of the property. But there is no law about how close people can live to the factory's property. They can live right on the borderline if they like."[5] Dr. Li says that, after Bhopal, government inspectors instructed 41 of the island's then 69 formulators to install a "poison-leaking warning system." (Three plants were shut down entirely.) In addition, "the government now is trying to find a new place for the pesticide manufacturers."

From his perspective as a researcher, Dr. Li says he thinks "pesticides are very dangerous, but if we handle them carefully they make a good contribution to agriculture." He feels it is his job "to minimize pesticide use here. The market is saturated in Taiwan." But pest problems require pesticides on some crops more than on others. "We have a good biological control system on sugar cane, but if we don't spray paddy crop, we lose 15.4 percent on the first harvest and 29.9 percent on the second harvest to pests."

Dr. Li also weighs the impacts of pesticides from the occupational health point of view. "We monitor urine samples from three groups: public service workers, farmers, and pesticide formulation workers. The first two are OK, but the [pesticide] workers show evidence of slightly elevated levels of the metabolites of organophosphates. The government forces the companies to check the blood and urine of workers every month. Many of them take a vitamin complex hoping to stimulate an enzyme that detoxifies some pesticides—this is a preventive step.

"There have been 30 suicides from paraquat in the past two years," Dr. Li says. "An emergency poison treatment center at Veteran's Hospital has computerized information about pesticides and antidotes to help other hospitals anywhere in the country when they get a case. From organophosphates and carbamates, poisoning victims can recover easily if treatment with atropine is prompt.

"Synthetic pyrethroids are safer," Dr. Li explains, "but it is easy for insects to become resistant to them. And once they are resistant, insects will stay resistant for a long time. I don't think they will lose that ability." Dr. Li says there is "a big problem with non-target [i.e., non-pest] organisms" getting hit by pesticides in Taiwan.

Unlike some Third World countries, Taiwan's pesticide regulations are actually tighter than those in the United States in some respects, including many of the food "tolerance" limits, which define how much residue is allowable in food sold to the public. In addition, Taiwanese women's breast milk is relatively uncontaminated compared to American women's, since the entire DDT-class of persistent chlorinated hydrocarbons is now completely banned and was not used in Taiwan as long as it was in the United States.

Despite its controls, Taiwan faces the danger of a pesticide-related disaster primarily due to the proximity of neighborhoods to the formulation plants. While Dr. Li does not minimize this danger, he does find it curious that related, everyday dangers from other chemicals are not more closely regulated as well. "In hospitals, for example, the same phenol compounds we worry about in pesticide plants are used to disinfect, but nobody says anything because they think of hospitals as safe places. And the government has not done anything to educate the housewife about the safe use of pesticides in the home. For instance, what about the possibility of long-term central nervous system damage from mosquito coils left burning all the time indoors? We also have no standard to regulate the smell of chemicals. If a household chemical smells nice, people think it is safe. But the smell is just perfume that the manufacturer added. Toxic products do not necessarily look or smell bad, but people just don't understand this."

While it is beyond the scope of this book to consider the wide variety of ways chemicals formerly confined to agricultural use have spread throughout modern society, Dr. Li's point is a good one. Even the best regulation inside factories will fail to protect people if consumers continue to use chemical products (and governments continue to regulate them) as if there will be no long-term consequences. Despite the hazards of immediate death as in Bhopal, the main danger presented by pesticides is more likely to be a gradual one, damaging genetic structures, for instance, or disrupting the essential health of the soil, water, flora, and fauna we depend on for sustenance.

Meanwhile, evidence of growing environmental awareness in Taiwan surfaced in 1986, when residents of the town of Lukang staged protests against plans by DuPont to build a $160 million chemical plant there. In response, *Business Week* reported, "The company is preparing an environmental-impact report and is blitzing Taiwan with information about the new, low-pollution technology planned for the plant."[6]

Notes

1. Various press clippings translated from the Chinese by Cassie Tao, June 1985.

2. Personal interview with the author, Taichung, Taiwan, June 8, 1985.

3. Personal interview with the author, Taichung, Taiwan, June 7, 1985.

4. Personal interview with the author, Taichung, Taiwan, June 7, 1985.

5. Personal interview with the author, Taichung, Taiwan, June 8, 1985.

6. "Suddenly, Asians are Fighting to Protect the Environment," *Business Week*, July 28, 1986, p. 23.

TEN

—

From
the Nile
to Mexico

It becomes good politics to let sleeping dogs lie.
Newspaper article, Zimbabwe

If methyl isocyanate were the only—or even one of the most important—intermediate chemicals used in pesticide production, future Bhopals might be quite easily avoided. MIC could simply be banned, or placed under the strictest possible controls, thereby considerably lessening the chance of another Bhopal-type disaster. But MIC is only one of many intermediates, most of which have never been studied carefully for their potential to cause disaster. "One of the ironies of the worst industrial accident in history," notes *C&EN*, "is that it involved a compound . . . that is far from being one of the mainstays of the chemical industry."[1] According to industry sources, the total amount of MIC used in 1980 was 38 million pounds, a mere 1.5 percent of the total estimated consumption of intermediates, which was about 2.5 billion pounds. At least 17 other intermediates are used in greater quantities than MIC. The largest group is the phosphorous compounds, which account for about 10 percent of the market, followed by alkylamines (precursors of carbamate pesticides), which account for about 8 percent. The phosgene derivatives, including MIC, are responsible for about 3.5 percent of the total world use of intermediates.[2] (See Appendix B, page 186.)

The world's ignorance about the hazards of intermediates extends to finished pesticide products as well. In 1983, the prestigious U.S. National Academy of Sciences reviewed the available data from the Environmental Protection Agency on 3,350 pesticides and their ingredients, and found that there was adequate information on the potential health hazards for only 10 percent of them.[3] Sometimes the reasons for this data gap are appallingly simple, as with MIC: "Because [MIC's] physical, chemical, and biological properties make it troublesome and unpleasant to work with," reports *C&EN*, "researchers have shied away from experimenting with it."[4]

The complex chemical reactions that occur inside factory reactors are not always well understood either. It is therefore extremely difficult for scientists to determine exactly how many controls are necessary to ensure an adequate degree of safety. As author Charles Perrow has demonstrated in his book, *Normal Accidents: Living with High-Risk Technologies*, it is possible that a series of small failures can set off unanticipated chemical interactions that even highly trained plant operators may not be able to interpret and control. "In the chemical industry, with mature technologies, good management, and strong incentives to keep the plants from blowing up, the persistence of systems accidents should suggest something intrinsic to the processes themselves," Perrow writes. Chemical plants are similar to nuclear power plants, where baffling and complex events in unpredictable sequences can lead to accidents like the one at Three Mile Island in 1979. Although accurate data are difficult to come by, Perrow quotes statistics that reveal an average of one and a quarter fires per year at large petrochemical refineries in the United States. He also states that the overall incidence of "systems accidents" appears to be high in chemical plants.[5]

In underdeveloped countries, where there's a lack of trained personnel, government standards, and other features of "industrial culture" to provide the necessary controls, chemical plants may represent potential time bombs. Particularly when a production facility turns into a losing proposition, as did the Bhopal plant, parent companies may be reluctant to invest the money needed to maintain the highest possible degree of safety. The decision to turn off the refrigeration unit at the Bhopal plant months before the accident was apparently such a cost-saving step.

Furthermore, when the 17-year international patent expires on a pesticide, small companies often begin manufacturing it. Most of Taiwan's producers, for instance, specialize in such "off-patent" products. The problem, from a safety point of view, is that the small producers may not maintain safety standards as high as those of the multinationals, which have larger resources and obligations to insurance companies and risk-management ombudsmen.

Despite the dangers, the perceived need for pesticides is so great in the underdeveloped nations that plants, both big and small, are being built practically everywhere. For this report, we surveyed knowledgeable people in Asia, Africa, and South America about pesticide plants in the areas where they lived or had visited. "I have seen some plants myself that would not be acceptable in our part of the world," said Jan Huismans, of the United Nations Environment Program. "In Dar es Salaam, Tanzania, for example, I saw a formulation plant under contract to a multinational where the workers were completely unprotected. The environment around the plant was exposed to dust and fumes, and the plant simply had no treatment facility for water discharge."[6]

And in Egypt, a prominent scientist reported that

a former dye company scheduled to start pesticide production in 1985 did not have "enough safety measures for pesticides." The wastes will be discharged from a drain into the Mediterranean Sea, in the vicinity of a "very crowded and highly populated area." He also identified a pesticide formulation plant near Cairo that had made "many discharges" into the River Nile, killing fish and causing conflicts with the surrounding "heavily populated areas."[7] An official from the Ministry of Agriculture in Cairo confirmed that the plant represented a "potential Bhopal."[8]

In Zimbabwe, a plant near Harare formulating a wide range of pesticides has come under criticism from people living nearby. "Since the Bhopal disaster, many residents living close to [the plant] have complained of a whitish gray cloud of sulfur dioxide fumes, and the plant's storage of ammonia," reported a local investigator.[9]

A professor in Liberia stated that one pesticide plant near Monrovia "is located in a very densely populated residential and industrial area, with very heavy, slow-moving traffic of cars and pedestrians at all times. . . . Should there be [an accident] the consequence to people and properties adjacent could be catastrophic."[10]

In Central America, pesticide manufacturing zones exist in the Dominican Republic and Guatemala. One investigator reported that most of Guatemala's pesticide plants "are practically in the middle of the city and we have no safety measures."[11] In Mexico, Ivan Restrepo Fernandez of the Center for Eco-development investigated conditions at pesticide plants and found many serious problems. "In our study," he wrote, "we discovered that many plants do not use protective equipment; and medical personnel are not available who could attend to workers who have been poisoned. When equipment does exist, it's neither adequate nor sufficient."[12] Although government inspectors levied fines on some operators for these deficiencies, the study found that companies considered it cheaper to pay the fines than to make the necessary repairs or adjustments to their equipment.

Some of the plants discussed in the study presented hazards to nearby communities. In one case, says Fernandez, "the principal of a school located across from a plant . . . asked health authorities to intervene and move the plant to a safer place, as the plant was a threat to the students as well as the general population. There had been a constant emission of gases and toxic substances [from the plant]. Three days after the principal's request, there was an explosion at the plant, which caused leaking of gas and fumes that contaminated the environment and the food supply of the local population."

In Malaysia, on March 1, 1986, a worker was reported to have died after inhaling chlorine leaking from a valve at one of the four plants in a large complex operated by the Chemical Company of Malaysia (CCM), a subsidiary of ICI. Besides the chlorine unit at the complex, CCM has a fertilizer plant, an animal feed plant, and a plant that produces paraquat. The complex has been a cause of controversy for years among nearby residents, 20,000 of whom live on the rubber and oil palm estates or in various *kampungs* ("villages") around Shah Alam, Selangor, near the capital city of Kuala Lumpur. According to environmental groups in Malaysia, villagers and students at a school close to the complex have difficulty breathing when dust, smoke, and gases vented from the stacks settle near the ground. The works manager of CCM was quoted in a press report, saying that the company's position was to meet with residents about any complaints, because it was company policy to be "good neighbors."[13]

The kinds of problems caused by chemical plants throughout the underdeveloped world suggest that, in their rush to industrialize, governments are skipping key stages in the establishment of what is called "industrial culture"—the web of institutions that developed almost simultaneously with industry in Europe and North America. "The technology, the government, the academic disciplines, the testing institutions, the technical personnel, and the insurance companies all developed together," says the ILO's Georg Kliesch. "They all have their special tasks. This is a check that produces an environment in which management is more willing to do what it should do. There is no doubt there is a lack of this sort of thing in the Third World."[14]

And a dangerous element of the accounts from all over the underdeveloped world is the location of plants in populated areas. There are many reasons why this occurs. Most of the services and infrastructure, including roads, electricity, public transportation, and supplies are located only in cities. The rural parts of underdeveloped countries remain largely undeveloped.

"Third World countries are accused of not taking sufficient care in this respect," notes one commentator. "Yet, Third World countries face particularly difficult social problems of zoning in their town plans. When a huge chemical plant is located away from the city precincts, thousands of workers who may be employed there face transport difficulties. Such workers often resort to building illegal shanty dwelling places around the chemical plant. Third World leaders have found it unpopular to attempt to destroy shantytowns mushrooming around hazard-ridden work places. It becomes good politics to let sleeping dogs lie."[15]

Notes

1. Ron Dagani, "Data on MIC's Toxicity Are Scant, Leave Much to be Learned," *Chemical & Engineering News*, Feb. 11, 1985.

2. Confidential pesticide industry reports.

3. Robert F. Wasserstrom and Richard Wiles, *Field Duty: U.S. Farmworkers and Pesticide Safety*, World Resources Institute, Study 3 (Washington, D.C.: July 1985), p. 24.

4. Confidential pesticide industry reports.

5. Charles Perrow, *Normal Accidents: Living with High-Risk Technologies* (New York: Basic Books, 1984).

6. Personal interview with the author, Geneva, Switzerland, Jan. 23, 1985.

7. Confidential communication with the author, May 1985.

8. Confidential communication with the author, May 1985.

9. Confidential communication with the author, May 1985.

10. Confidential communication with the author, May 1985.

11. Confidential interview with Jonathan King, Guatemala City, Guatemala, Mar. 1985.

12. Personal communication with the author, Mar. 4, 1985.

13. "Worker Dies After Inhaling Chlorine Gas in Fac-

tory," *Suara* SAM (Penang, Malaysia), May 1986, pp. 2–3.

14. Personal interview with the author, Geneva, Switzerland, Jan. 24, 1985.

15. Undated press clipping, Zimbabwe.

ELEVEN

Worrying at Kurosaki

I always carry a beeper in my back pocket, and I always worry, even when I'm out having a drink with the men.
Official, Mitsubishi Chemical Industries Ltd.

The Kurosaki plant of Mitsubishi Chemical Industries sprawls over 815 acres along Dokai Bay in the northern Kyushu industrial belt, just across the railroad tracks from the busy downtown section of Kitakyushu, one of Japan's major port cities. As a main iron and steel production center during World War II, this industrial area was the original target for the atomic bomb that, due to bad weather over Kitakyushu, was dropped on Nagasaki instead.

A fairly strong chemical smell from the plant can be detected in the downtown area, but residents say it used to be much worse. Water pollution in the area also had reached unacceptable proportions, partly due to the plant's location on landfill that contains industrial waste materials. But both water and air have been cleaned up in recent years.

The Mitsubishi plant manufactures pesticides (and other products) for both domestic and export markets, and is the only one in Japan that uses MIC. Kosaku Arie, chief of the general affairs division of the Kurosaki plant, explains why he thinks a Bhopal-type catastrophe won't happen here:

"Our focus is on preventive maintenance. We maintain a 200-member team to check if anything goes wrong anywhere. Our checking system is computerized and it checks the pressure and temperature of each section of the process. It shuts down production automatically whenever there is a problem.[1]

"Water never gets into our system. The MIC plant is inside a building which maintains lower pressure than the atmosphere outside. The phosgene [used to make MIC] also has only on-site production."

Unlike Union Carbide, Mitsubishi's "closed loop" system produces phosgene and MIC only in the amounts needed for immediate conversion into carbamate pesticides. The company says no more than about 250 gallons of MIC exist at any one time in the plant.

Still, Arie worries about the possibility of a runaway reaction from MIC or one of the 49 other hazardous chemicals at the plant. "If an accident ever *should* happen we would lose our company's good name, and that's always on our mind," he explains. "Our employee housing is near the plant, so we are worried about ourselves. I always carry a beeper in my back pocket and I always worry even when I'm out having a drink with the men."

The day after the Bhopal disaster, an inspection team from the Kitakyushu Fire Department showed up at the plant to tour the MIC unit. Tetsu Ito, chief of the guidance division for the fire department, says that the company "meets all our standards. Our only recommendation was to heighten all safety consciousness." Ito adds, however, that Mitsubishi is obligated to report a leak to his department "only in the event of a fire."[2]

Officials from the Labor Standards Office in Japan also admitted that there was "no rule to report a leak or spill of phosgene or MIC as long as there's no human impact." The officials added that they conducted their first inspection ever of the MIC unit after Bhopal and found it well-maintained.[3] (In addition to these government inspectors, a coalition of citizens' organizations, the Bhopal Disaster Monitoring Group, requested to inspect the MIC unit, but Mitsubishi refused to allow them inside.[4])

For most of the past decade, the Mitsubishi operation appears to have had a good safety record. During the previous decade, however, the aged facility, which has since been upgraded, was plagued with problems. One employee died and five were injured in a cyclohexylamine tank explosion in 1967; sulfuric anhydride enveloped the local railway station in a chemical mist after an accident in 1969; an explosion of high pressure hydrogen occurred at the ammonia plant in 1970; 62 employees were made ill by phosgene leaking from a pipe in 1971; 23 workers were hospitalized due to exposure to chlorine after an explosion in 1973; about 1,000 liters of inflammable coal tar leaked in 1975; and one employee was killed and six were injured by a boiler explosion at the ammonia plant in 1976. Since that series of accidents, however, according to the Labor Standards Office, the plant has maintained a safety record better than that of the chemical industry as a whole in Japan.[5]

Although Mitsubishi has no MIC plants overseas, according to Kosaku Arie, "if we built an overseas plant, it should be as good as or better than ours here, because we know what's wrong with our plant and should be able to improve it over there. There should not be a double standard for Third World countries."

Unfortunately, Arie's opinion is not necessarily shared or wholly practiced by Japanese industry. There are numerous examples of Japanese manufacturing plants in underdeveloped countries that fit the category of "runaway hazardous industries." Mitsubishi's subsidiary, Malaysian Rare Earth Corp. Sdn. Bhd., for instance, produces a radioactive waste product, thorium hydroxide, in the process of manufacturing yttrium oxide concentrate, and engages in waste disposal practices that have drawn criticism in Malaysia.[6] And a subsidiary of the Asahi Glass Company was found to be dumping organic mercury—the cause of Japan's terrible Minamata disaster—into one of Thailand's most important rivers.[7]

It is practices such as these that fuel the growing international controversy over double standards and runaway hazardous shops. European companies in particular have been among the most aggressive in penetrating Third World markets, sometimes producing products that are no longer permitted for use back home. The Royal Dutch Shell Group, for example, which ranks as one of the world's largest corporate empires and is the fourth-largest pesticide producer, manufactures the highly restricted chlorinated hydrocarbon pesticides aldrin and dieldrin at plants in both the Dominican Republic and the Sudan.[8] And the global pesticide map of Germany's Hoechst reads like a "Who's Who" of banned and heavily restricted pesticides: toxaphene in Zambia and Thailand; chlordane in Zambia and Ecuador; endrin in Thailand; DDT in Zambia, Thailand, and Brazil; heptachlor in Thailand; HCH in Zambia; 2,4,5-T in Peru; 2,4-D in Thailand; lindane and the Bhopal pesticide, carbaryl, in the Sudan; disulfotane in Thailand; endosulfan in Malaysia, the Philippines, Thailand, Colombia, and

Peru; methamidophos in Indonesia, Thailand, and Peru; paraquat in Peru, Thailand, and Colombia; and parathion in Colombia and the Philippines.[9]

The flow of pesticide technology from the developed to the underdeveloped world is thus well established. But, upon closer examination, how safe is that technology back in the largest country of origin—the USA?

Notes

1. Personal interview with the author, Kitakyushu, Japan, June 5, 1985.

2. Personal interview with the author, Kitakyushu, Japan, June 5, 1985.

3. Personal interview with Jun Ui, Kitakyushu, Japan, June 5, 1985.

4. Personal correspondence from Nobuo Matsouka with the author, Sept. 27, 1985.

5. Based on research by the People's Research Institute on Energy and Environment (PRIEE), Tokyo, Japan, June 1985.

6. IOCU press file, Penang, Malaysia.

7. Research by PRIEE, June 1985.

8. Dr. Antonio Thomen, "Los plaguicidas en la Republica Dominicana," *Ahora!* no. 1063, Apr. 9, 1984.

9. Wood, Mackenzie & Co., 1984; see also "Deutsches Gift in Aller Welt," Pestizid Aktions Netzwerk (Hamburg), Dec. 1984.

TWELVE

It Can
Happen
in America

To operate that plant without any leaks at all for any
length of time is just beyond our capabilities.

Union Carbide official

The world's biggest pesticide producer and the biggest pesticide user is the United States. Since Bhopal,
greater public and media attention has been focused
on the safety of this giant industry, which accounts
for roughly a third of the global market every year.
And although the industry claims it is one of the safest
in the country, it has been plagued with a series of
embarrassing accidents just at a time when the public
spotlight is at its brightest.

Most Americans probably still believe that a
Bhopal-type accident could not happen here, that the
U.S. plants are somehow safer, the laws more effective, and the political leaders more responsible than
in Third World countries. But most Americans also

live in a state of near ignorance about the chemical factories in their own communities. According to a 1985 report by the Congressional Research Service, about 75 percent of the U.S. population lives "in proximity to a chemical plant."[1]

Just two weeks before the Bhopal disaster, residents of Middleport, New York, discovered the dangers posed by an FMC Corporation pesticide plant 1,000 yards from a grammar school in their town. In November 1984, what FMC initially called a "raw material"—used to produce their pesticide *Furadan*—spilled at the plant during a transfer of the chemical from a tank to a processing unit. Vapors from the spill entered the school's ventilation system, leading to the evacuation of 500 children, and sending nine of them, as well as two teachers, to the hospital with eye irritation and respiratory difficulties. The chemical turned out to be MIC.[2] The city fire chief complained about the "lapse of time between the actual spill and notification of outside support agencies, school authorities, ambulance, fire, and evacuation personnel."[3]

It wasn't until the disaster in Bhopal that Middleport residents realized that a lethal chemical was being routinely shipped and handled in their community. Diane Hemingway, a mother of two children who attended the school, polled parents and found that many of their children complained of lingering health effects, including headaches, skin rashes, and coldlike symptoms. Hemingway's son, who had asthma symptoms before the accident, has since suffered several full-blown asthma attacks.[4]

A group of parents demanded that physical examinations be conducted to determine possible long-term effects on their children; the examinations were finally carried out in June 1985, seven months after the spill. Partly due to the elapsed time and the difficulty in calculating exposures, the tests proved inconclusive.[5]

FMC, which provides 170 jobs and 20 percent of Middleport's tax base, has retained widespread support for its plant from residents of the economically depressed area. When the plant was temporarily shut down and workers laid off after the Bhopal disaster, many worried that the plant would not reopen. "It'll be like Christmas if we ever see a train car full of MIC pulling in here," David Bilicki, an FMC shop steward, told the *New York Times*.[6]

After inspecting the plant, the U.S. Occupational Safety and Health Administration (OSHA) found FMC guilty of four "serious violations" and labeled it the most seriously deficient of all five U.S. plants using MIC. According to OSHA, the plant did not have the necessary safety systems to prevent an uncontrolled chemical reaction of MIC like the one that led to the Bhopal tragedy, though OSHA considered the chance of such an incident here "extremely remote." OSHA also reported that five "recorded leakages" of MIC had occurred at the plant since 1982.[7]

Since the spill, FMC has installed new safety measures at the plant. But Hemingway is still worried, not only about the threat of a major accident but also about the long-term health effects of exposure to low-level emissions of MIC. New York's Department of Environmental Conservation says the plant releases a small amount of MIC all year round during the course of its normal operations.[8] Accordingly, Hemingway and her family recently relocated to another town in the area, and she moved the children to a different school.

The month before the FMC spill, on October 6, 1984, a letter carrier in a small town in New Jersey felt his eyes begin to burn and his breathing tighten. In nearby Linden, at American Cyanamid's pesticide factory, a tank containing 12,000 gallons of crude malathion had overheated, bursting a valve. The resulting fumes extended for 20 miles across at least seven communities in New Jersey and neighboring New York. Over 100 people were taken to hospitals for treatment of chemical poisoning symptoms like nausea and dizziness.[9]

The malathion leak was only one of a score of chemical accidents, averaging almost one a week, from October 1984 through January 1985, in New

Jersey's heavily industrialized Union and Middlesex counties. Several of the accidents occurred at Cyanamid's plant; there were others at plants owned by DuPont, Exxon, Union Carbide, and other companies.[10] The state's Department of Environmental Protection (DEP) fined Cyanamid and DuPont (which together were responsible for 50 percent of the accidents), and ordered them to conduct plantwide safety evaluations.[11]

"We're concerned about the number of accidents," says DEP assistant administrator George Tyler. "We have an aging industrial base and decisions may have been made to go easy on maintenance at some of these plants. If you are going to phase out a plant, you are going to do what you can with chicken wire and chewing gum rather than investing in maintenance and equipment. That's just a suspicion on my part." But Peggy Ballman, assistant director of the Chemical Industry Council of New Jersey, rejects Tyler's hypothesis. "We know these facilities," says Ballman. "They are using modern equipment. The buildings may be old but the equipment is state-of-the-art."[12] If Ballman is correct, the string of accidents is evidence that the technology itself is inherently dangerous, and uncontrollable by even the latest and best safety equipment.

Dramatic evidence of this occurred in August 1985 at Institute, West Virginia, the site of Union

Carbide's *other* MIC plant. After the Bhopal tragedy, Union Carbide had shut down its Institute plant and installed an elaborate $5 million computerized leak detection and early warning system known as Safer. The new system is supposed to identify instantly the speed and direction of chemical leaks into the atmosphere, thereby giving plant operators time to warn and evacuate nearby communities. Nevertheless, on August 11, 1985, the system failed. A cloud of mixed chemicals, aldicarb oxime and methylene chloride, used in the manufacture of aldicarb (*Temik*), escaped when a gasket failed in a 500-gallon storage tank. According to a Union Carbide spokesperson, aldicarb oxime may contain trace amounts of MIC. Methylene chloride is toxic to the nervous system and is a suspected carcinogen under special EPA review.[13]

As the cream-colored chemical cloud spread over four neighboring communities, residents were confused about what they should do. After Bhopal, they had been assured by Union Carbide that any leak that endangered their health would be detected in time to warn them with emergency sirens. But for 20 critical minutes, according to residents, no siren sounded. Only later did they find out why. The sophisticated early warning system had indicated to plant operators that the cloud was hovering over the plant, when it was actually settling onto surrounding communities.

Union Carbide officials had failed to program the computer system to detect the *specific chemical combination* that leaked, so it was of no help in the emergency. In fact, by installing the Safer System but not programming it correctly, Union Carbide officials may have created a false sense of security among operators who were depending on the computer to help them make decisions. By the time they realized their error, thousands of people had been exposed to the chemical cloud and over 130 were headed for hospitals with burning sensations in the eyes, nose, throat, and lungs.

In addition to lingering health effects, officials warned that food grown in the soil near the Institute plant might be contaminated and should not be eaten by residents, who are mostly poor, working-class people. OSHA later cited Union Carbide for "willful neglect" of numerous safety procedures at the plant, and proposed the maximum fine of $32,100 as the penalty.[14]

The danger to workers in U.S. pesticide plants was demonstrated earlier in 1985 when James Cox, an employee at Vertac Chemical's plant in Jacksonville, Arkansas, died after two to three gallons of liquid intermediate phenols used to make 2,4-D splashed on him during a sampling procedure. The chemical burns covering 75 percent of his body were from a chemical mixture so potent that a paramedic treating Cox was burned also. Four years earlier, another worker had died of chemical burns at the same plant.[15]

Other recent accidents have affected those who live near U.S. pesticide plants. On June 22, 1985, about 10,000 people had to be evacuated from the southern California cities of Anaheim (home of Disneyland), Fullerton, and Placentia, when toxic fumes escaped from a chemical fire at a pesticide warehouse. Those affected complained of sore throats, dizziness, breathing difficulties, and eye irritations.[16] Four days later, a warehouse storing 25 tons of pesticides caught fire and exploded in Coachilla, California, leading to the evacuation of 2,000 people from a nine-square-mile area. Some were treated at hospitals for cramps, nausea, and respiratory ailments.[17]

Some subsidiaries of U.S. companies operating in other developed countries also have had recent accidents. For instance, in New Zealand, Ivon Watkins-Dow, reportedly the world's only remaining producer of the notorious herbicide 2,4,5-T, accidently released over 3,000 pounds of sodium trichlorophenate into the air at its New Plymouth plant, including an unknown amount of deadly dioxin, on April 5, 1986. The company is majority owned by Dow Chemical, and is the object of growing public controversy in New Zealand.[18]

In the United States and overseas, the accidents

continue, usually without fatalities, but often many people in surrounding communities are exposed to the chemicals. The long-term impact of pesticide poisoning on these victims' health is unknown. And the possibility of a worse accident remains ever present.

"I think you're going to have to live with some of it [risk]," Union Carbide's vice president for health, safety, and environmental affairs, Jackson Browning, told a congressional committee. "Nobody wants a leak to occur. But to operate [the Institute] plant without any leaks at all for any length of time is just beyond our capabilities."[19]

"There is no question that 'Bhopal' could happen in the U.S.," says Richard Boggs, a management consultant on employee safety and health issues. "You are dealing with such terribly dangerous chemicals that human failures or mechanical failures can be catastrophic. The potential is here and it could happen, maybe today, maybe 50 years from now."[20]

Notes

1. James McCarthy, Congressional Research Service Report (Washington, D.C.: Feb. 22, 1985).
2. Testimony before the U.S. Senate Committee on En-

vironment and Public Works (Newark, N.J., Feb. 18, 1985).

3. *Journal-Register* (Medina, N.Y.), Dec. 5, 1984.

4. Telephone interviews with Connie Matthiessen, CIR, Washington, D.C., July 9 and 15, 1985.

5. "MIC Tests Completed," *Journal-Register*, June 20, 1985.

6. "Village's 'Heartbeat,' A Chemical Plant Raises Fears," *New York Times*, Mar. 9, 1985.

7. Douglas Turner, "U.S. Agency Raps FMC's Handling of Deadly Chemical," *Buffalo News* (N.Y.), Feb. 16, 1985.

8. Richard Moss, "MIC Regularly Released into the Area Atmosphere," *Journal-Register*, Dec. 18, 1984.

9. Testimony, Committee on Environment and Public Works; see also Barry Meier, "Bhopal Spurs U.S. Scrutiny of Chemicals," *Asian Wall Street Journal*, June 14–15, 1985.

10. Lisa Peterson, "Chemical Plant Mishaps Fuel Concern in Jersey," *Star-Ledger* (Newark, N.J.), Jan. 14, 1985.

11. "New Jersey's Crackdown on Toxic Emissions," *Chemical Week*, Mar. 13, 1985.

12. Telephone interviews with Connie Matthiessen, CIR, Washington, D.C., Sept. 17, 1985.

13. "Toxic Cloud Leaks at Carbide Plant in West Virginia," *New York Times*, Aug. 12, 1985.

14. Associated Press, "Carbide Admits Delay in Alerting Officials of Leak," *Ft. Myers News Press* (Fla.), Aug. 13, 1985; "Union Carbide Halts Poison Gas Production," Aug. 14, 1985; "Officials Warn Carbide Leak Poses Lin-

gering Danger," Aug. 15, 1985; "Gas Cloud More Toxic Than First Thought," Aug. 16, 1985; see also Kenneth Noble, "U.S. Seeks to Fine Carbide Over Leak," *New York Times*, Oct. 2, 1985.

15. "Vertac Worker Killed in Toxic Chemical Spill," *Arkansas Democrat*, Apr. 17, 1985.

16. Marcida Dodson, "Chemicals Found 'Haphazardly Stored' at Fire Site," *Los Angeles Times*, June 26, 1985.

17. Louis Sahagun and Jerry Belcher, "Fire-caused Toxic Smoke Routs 2,000 in Coachilla Valley," *Los Angeles Times*, June 27, 1985.

18. "Badly Fitted Disc Blamed for Leak," *Taranaki Herald* (New Zealand), June 17, 1986.

19. United Press International, "Union Carbide Says It Can't Prevent Leaks," *San Francisco Chronicle*, Feb. 20, 1985.

20. Telephone interview with Connie Matthiessen, CIR, Washington, D.C., Sept. 17, 1985.

PART THREE

The Solution

THIRTEEN

Power
and
Money

It is not proper for an international corporation to put
the welfare of any country in which it does business
above that of any other.

Union Carbide spokesperson[1]

When the warning sirens sounded, belatedly, at
Union Carbide's MIC plant in Bhopal on December 3,
1984, some of the people living nearby awoke and
started running *toward* the plant, thinking it was on
fire.[2] Unaware of the inherent dangers, they were simply trying to be good neighbors and to help put out
the fire.

The company work force, however, all too aware
of what was happening, ran in the opposite direction—*away* from the crippled plant. As they ran, they
passed four buses parked on the factory grounds that
were part of a contingency for evacuating nearby residents in an emergency.[3] The buses were never used.

The contrast between the multinationals and their host countries is startling. Union Carbide was able to achieve sales of over $86,000 per employee in a recent year, while the per capita GNP for India is only $260.[4] Economists would be quick to point out that these yardsticks measure different phenomena—that to compare them is like mixing apples and oranges. Indeed, the multinationals pledged to maximize profits for their shareholders, and the Third World governments trying to join 20th century industrial society are very different types of entities.

The complex relationship between the multinationals and their host governments was perhaps best described by Richard J. Barnet and Ronald E. Müller in their seminal study, *Global Reach*: "The survival of the poor nations of Asia, Africa, and Latin America now depends upon how and on what terms they can relate to the world industrial system."[5] This dependence leads to a situation where "the three essential structures of power in underdeveloped societies are typically in the hands of global corporations: the control of technology, the control of finance capital, and the control of marketing and the dissemination of ideas."[6]

The view of those sympathetic to the expansion of the multinationals was well stated by George Ball, former U.S. under-secretary of state and chairman of Lehman Brothers International: "Working through great corporations that straddle the earth, men are able for the first time to utilize world resources with an efficiency dictated by the objective logic of profit." By contrast, said Ball, the nation-state "is a very old-fashioned idea and badly adapted to our present complex world."[7]

But the relationship between corporations and Third World governments is a two-way dependence. "The resources of the poor nations, including the raw materials on and under their territory, cheap labor represented by their teeming populations, and the potential customers represented by their expanding middle classes are . . . increasingly crucial to the plans of the global corporations," Barnet and Müller write. "Incredible as it may seem, the poor countries have been an indispensable source of finance capital for the worldwide expansion of global corporations."[8] For example, according to the Bank for International Settlements in Switzerland, the Third World provided twice as much money as it received from the international banking system in 1984.[9]

The relationship between the multinationals based in rich countries and the "underdeveloped" nations is such that, some have suggested, the adjective may actually be a verb—i.e., that Third World countries are being systematically *underdeveloped* rather than developed by the financial/industrial institutions of the United States, Europe, and Japan.[10]

In the aftermath of the Bhopal tragedy, certain hidden aspects of the complicated financial network underlying huge multinational investments in countries like India have suddenly become partially exposed. The insurance industry, for example, which flourishes in relative obscurity compared to its chemical company customers, is deeply involved in the litigation surrounding the Bhopal case. Most multinationals maintain a basic layer of insurance against the possibilities of accidents, called a "self-insured retention." They then purchase a "primary layer" of general insurance from one of the big insurance companies. And then, on top of that, they typically buy several layers of "excess" insurance to cover whatever claims they feel they are likely to face from the public.[11] The enormity of the Bhopal catastrophe has apparently stretched the limits of this system, however. "Already," the trade journal *Business Insurance* reported three months after the accident, "the Bhopal incident is putting pressure on worldwide insurance capacity, and multinationals can expect a sharp boost in the premiums they are paying for coverage overseas."[12]

"Rates are skyrocketing at all levels," reported one insurance expert. "There has been a reduction of insurance capacity. You can't go out and buy a billion dollars' worth of coverage any more."[13] Chemical companies in particular are reportedly finding it difficult to obtain any "pollution insurance" at all. Soon, one industry expert predicted, "no pollution coverage will be available to anyone."[14] And, in a filing with the Securities and Exchange Commission at the end of the first quarter in 1985, Union Carbide revealed that it had been only able to obtain "substantially less insurance with broader exclusions and substantially larger deductible" than in 1984.[15]

The British press, which covers the insurance business closely, reported that a large international broker, Marsh & McLennan, arranged Union Carbide's "umbrella liability insurance," and that the company also insured the Bhopal plant with the state-owned Insurance Company of India.[16] Lloyds of London is reported to have reinsured many of the primary policies and as many as 60 insurance companies may be involved in Union Carbide's exposure to liability.[17] Union Carbide is believed to have at least $200 million in total coverage, but legal claims could run into the billions.

The long-term effects of Bhopal on the ability of chemical companies to obtain insurance may therefore prove devastating. The *Wall Street Journal* reported that the "pollution-liability insurance field has virtually collapsed"; and *Chemical Week* quoted a leading broker as saying that if large, "U.S.-style" damage awards result from the mountain of litigation following Bhopal, "it could blow the lid right off the international insurance market."[18] "For companies such as Union Carbide," stated *Business Week*, "there may never again be such a thing as enough insurance."[19]

The financial stakes, then, are high, pending decisions in courtrooms and bargaining councils over how much Union Carbide is obligated to pay victims of the disaster. The company may conceivably be driven into bankruptcy. Whether or not that happens, governments may be pressured to provide agrochemical producers with greater protection from liability. All over the world, corporate executives and government officials are watching anxiously to see whether the Bhopal case alters the terms of their relationship—or whether they will soon simply be able to get back to business as usual.

Notes

1. Richard J. Barnet and Ronald E. Müller, *Global Reach: The Power of the Multinational Corporations* (New York: Simon and Schuster, 1974), p. 16.

2. Stuart Diamond, "The Pain of Progress Racks the Third World," *New York Times*, Dec. 9, 1984.

3. Ibid.

4. The World Bank, *World Development Report* (New York: Oxford University Press, July 1984).

5. Barnet and Müller, *Global Reach*, p. 136.

6. Ibid., p. 146.

7. Ibid., pp. 13–14, 19.

8. Ibid., p. 125.

9. Agence France Presse, "More $$ Flow Out of Third World Than In," *Star*, Malaysia, May 1, 1985.

10. Various publications (San Francisco: Institute for Food and Development Policy).

11. Confidential interviews with the author.

12. S. Robert Beane, "Reaction to Bhopal," *Business Insurance*, Mar. 4, 1985.

13. "Insurance Coverage for Chemical Firms Will Be Less Available, More Restrictive," *Chemical & Engineering News*, Feb. 11, 1985, p. 48.

14. Daniel Katzenberg, "Liability Insurance: The Added Burden of Bhopal," *Chemical Week*, Feb. 13, 1985, pp. 30–31.

15. "Chemical Maker Finds it Hard to Get Insurance," *Wall Street Journal*, Apr. 2, 1985.

16. Bailey Morris, "Carbide Can Pay All Claims," London *Times*, Dec. 8, 1984.

17. Terry Dodsworth, "Shadowy World of Multinational Disaster Insurance," *Financial Times*, Dec. 19, 1984.

18. Katzenberg, "Liability Insurance," pp. 30–31; see also Mary Williams Walsh, "Risky Business: Insurers Are Shunning Coverage of Chemical and Other Pollution," *Wall Street Journal*, Mar. 19, 1985.

19. "Carbide's Coverage: An Ironic Footnote," *Business Week*, Dec. 13, 1984, p. 40.

FOURTEEN

Our Obligation to History

People must have the right to know.
Robert Engler, author

In the course of this brief inquiry, we have seen that multinational corporations producing pesticides and other hazardous substances have built plants around the world; and that small, local companies making the same products, often with less rigid controls, have started up in many places as well. We've seen that pollution from these plants is difficult to control, and that chemical emissions are fouling the air and water from the Mediterranean to Rio de Janeiro. We've seen that leaks and spills plague some of these factories, even those in developed countries with the latest in safety technology, and that many industry insiders are worried that bigger accidents might someday occur. We've seen how changing market and field conditions for pesticides mean that new products are constantly being introduced, outpacing the ability of engineers and government regulators to institute adequate controls. Particularly in the Third World, we've seen an overall lack of control infrastructure that amounts to a "syndrome" of inadequacies, making accidents all the more likely there. We've seen that MIC—responsible for the worst industrial accident in history—is only one of many pesticide "intermediates," the health effects of which have scarcely been studied. This appalling data gap is not limited to the community level where chemical plants operate but reaches to the national and international levels, encompassing the agencies of the United Nations and the most prestigious scientific councils of the United States. And side by side with this global pesticide proliferation we find people—millions of people, mostly poor and uneducated—living right next door to potential catastrophe.

Considered as a whole, the situation seems ripe for disaster. A technology intended to help humankind appears to have gone berserk, beyond the point where we could hope to reduce the threat to a manageable proportion. But appearances can be deceiving, and solutions may lie closer within our grasp than is commonly realized.

In the wake of Bhopal, a simple but very powerful idea has begun to gain acceptance among the people who work in or live near chemical factories. Launched by U.S. organized labor in the early 1970s, it is called "right-to-know" and it means that those who may be affected by hazardous technologies have an inalienable claim to all relevant health and safety information to determine the nature of the threat, and what, if anything, they want to do about it. By September 1985, approximately 30 states and 50 communities had adopted right-to-know laws, and national legislation was being considered. The idea was gaining currency abroad as well.

"People must have the right to know what hazards science has created," writes American author Robert Engler, in *The Nation* magazine. "They didn't at Seveso, at Three Mile Island, at Love Canal. The immediate consequence of not knowing was that local health and fire departments could not prepare for the 'unlikely' or know what to do when the 'impossible' happened, as at Bhopal. Technological decisions that will affect communities should first be debated by the people who live there."[1]

Closely allied with the right-to-know movement is "freedom of information," the tool by which citizens and reporters can dig out relevant data about chemical issues and other matters of public concern from government files that normally remain inaccessible. Yet, it is disturbing that in the course of research for this book, many barriers to freedom of information about pesticides were encountered. In the United States, Europe, and Japan, figures for the production and sale of specific pesticides are routinely classified as "trade secrets" and hidden from the public. Health and safety data, if collected, often seem to overlap with "proprietary information" and therefore are not released either. In the Third World, leaking information about pesticides can cost conscientious officials their jobs. In some countries, anyone who reveals the results of government studies about the dangers of pesticide plants can be sent to jail. In one country, our investigators were warned that attempts to visit and photograph an American-owned pesticide plant would be extremely dangerous. In another, the involvement of a member of a prominent family in the ownership of a controversial pesticide factory was kept secret. Releasing that information, we were told, could lead to a governmental backlash that would drive concerned organizations out of existence and send their leaders to jail. In many Third World countries, reporters try to do their jobs under restrictions that their counterparts in the industrialized world would consider outright censorship and repression. The ability of unionists to organize workers and educate them about the dangers of the products they work with is similarly nonexistent in many parts of the globe. Even in the United States, longtime labor leader Tony Mazzocchi of the Workers Policy Project in New York says that workers "are scared to talk about the problems that these plants have. There is a lot of job blackmail going on."[2] In a time of high unemployment, companies routinely threaten to move operations elsewhere if workers or communities protest too loudly.

Right-to-know and freedom of information laws cannot solve these problems, but they do establish the framework for making technology more democratic. Armed with more accurate information, citizens can better judge the costs and benefits of allowing pesticide plants to operate in their communities. Thus informed, they are better prepared to participate in the decisions that can ultimately determine not only their economic well-being but the state of their health and that of their children.

The proximity of neighborhoods and schools to chemical plants around the world necessitates more immediate action, however, than the long-term educational efforts of the right-to-know movement. Given the potential for disaster, some experts believe that safety zones should be established around these facilities. The Bhopal tragedy added impetus to one such effort in France, where draft legislation has been introduced requiring companies constructing high-risk industrial facilities to compensate neighboring landowners for restrictions on the right to develop housing nearby.[3]

Many other forms of stringent restrictions may be necessary, including requirements that manufacturers supply complete, up-to-date information to health and emergency agencies, and that plant officials guarantee immediate access to health and safety inspectors making on-site investigations.

Bhopal may well be the spur for strong new regulations over the manufacturing of chemicals in many parts of the world. In the meantime, however, the concerned public and policymakers are trying to develop support for reducing the overall dependence on chemicals. In the case of pesticides, for example, there are many ways to grow food and control pests without relying on large amounts of chemicals. The use of biological controls—using natural enemies of pests—is one approach. Cultural controls, including crop rotation, interplanting with crops that repel pests, removing residues of old crops, and other traditional farming practices often prove successful. Physical controls that involve blocking or trapping pests are another set of alternatives. The use of resistant varieties of crops also thwarts unnecessary pest damage. All of these methods, plus the use of new, much less toxic chemical controls, represent significant alternatives to the current dependence on highly toxic pesticides. They are often grouped under the general category of "integrated pest management" (IPM), and rely on pest monitoring programs in the place of routine pesticide applications.[4] The success of IPM has been demonstrated in many places and on many major crops. A six-year joint study by three U.S. government agencies, for example, revealed that IPM could reduce the amount of pesticides used on the four leading U.S. crops by 70 to 80 percent in ten years' time, without reducing yields.[5]

It is overseas, in Third World countries, where the need to implement alternative pest-control strategies is most urgent. The World Health Organization has estimated that at least half of the world's acute poisonings take place in underdeveloped countries every year, even though they use only one-sixth of the world's pesticides.[6] In other words, people are being poisoned at a rate three times that of their counterparts in the industrialized countries. Every 42 seconds, somebody is poisoned somewhere, leading to at least ten thousand deaths annually from pesticides in the Third World alone.[7] Furthermore, at least half of all pesticides used in underdeveloped countries are applied, not to local food crops, but to luxury items (coffee, tea, bananas, etc.) and to nonfood crops (cotton, rubber) grown for export to the United States, Japan, and Europe.[8] Therefore, the major argument advanced in favor of Third World pesticide use—helping feed the hungry—is a highly questionable one.

Until alternative pest-control techniques are implemented on a large scale in modern agriculture, the market for the chemicals that cause Bhopal-type accidents will continue to expand. New plants will be erected to satisfy the growing markets, spreading the danger of additional catastrophes. The tragic irony of Bhopal is that the deaths and injuries were caused during the production of a pesticide not needed in India. Many argue that we have the knowledge to reduce, if not eliminate, the agricultural use of chemical pesticides altogether, but that our current economic and political systems encourage their continued use. Perhaps some future generation will look back upon our agricultural age as an extremely primitive one that relied on crude and self-defeating chemical weapons in a war that could not be won. Blinded by what amounts to a global chemical blizzard, we can't seem to find our way off the pesticide "treadmill." The world's leading food producer, the United States, produces 500 billion pounds of synthetic organic chemicals every year to only 50 billion pounds of food. In the words of American author and ecologist Barry Commoner, "We have produced a chemical system antagonistic to life, which is essentially overwhelming it."[9]

It will be difficult for citizens to challenge the entrenched and politically powerful chemical industry. But as Commoner has observed, the "social necessity" of that industry must be openly debated in the press and in the halls of government. Only by curbing the spread of agrochemicals around the world can the fierce momentum of "the Bhopal Syndrome" be slowed and eventually broken.

This process will be long, slow, and difficult. But growing numbers of people feel they owe it to themselves not to "abdicate," in Robert Engler's words, "our obligation to shape history."[10] That obligation is one each succeeding generation hopes to turn over one day to its children, and to the generations that, according to the natural order, should be assuming stewardship of the planet. The ability of future generations to meet their obligation, however, now hangs in the balance, depending on our ability to meet ours.

Notes

1. Robert Engler, "Many Bhopals: Technology Out of Control," *The Nation*, April 27, 1985.

2. Telephone interviews with Connie Matthiessen, CIR, Washington, D.C., July 15 and 22, 1985.

3. *Business International*, Mar. 8, 1985.

4. Toby Stewart, "Consider The Alternatives!" PAN International "Dirty Dozen" packet, June 5, 1985.

5. Ralph Brownstein and Ralph Nader, *Who's Poisoning*

America? (San Francisco: Sierra Club Books, 1981), p. 317.

6. David Bull, *A Growing Problem: Pesticides and the Third World Poor* (Oxford UK: Oxfam, 1982).

7. Ibid.

8. David Weir and Mark Schapiro, *Circle of Poison: Pesticides and People in a Hungry World* (San Francisco: Institute for Food and Development Policy, 1981).

9. Speech delivered at the conference "After Bhopal, Implications for Developed and Developing Nations," Newark, New Jersey, Mar. 21, 1985.

10. Engler, "Many Bhopals."

CONCLUSION

Path Toward a Future

Confronted with a group of signs and symptoms that collectively indicates a disease or abnormal condition, doctors sometimes use the term "syndrome," which comes from a Greek word meaning "a running together" (of symptoms). It is the convergence of many factors involved in the global pesticide business today that makes the Bhopal Syndrome an abnormal condition that has larger implications for the value of human life. Today's insecticides, herbicides, fungicides, plant growth regulators, and other synthetic agrochemicals are the products of human ingenuity, but also of human limitations. We have the ability to synthesize wonderfully complex, previously nonexistent chemical combinations that help reduce—at

147

least temporarily—the competition with other creatures for the food and fiber we grow in our fields. But our willingness to use these powerful weapons is based on an inadequate understanding of the intricacies of the finely balanced ecosystems in which we operate.

Perhaps the most enduring contribution to human progress made by pesticides is an ironic one, as Dr. Ross Hall of Environment Canada has suggested.[1] In the fields, chemical pesticides never kill 100 percent of a target species. Insects, which are the usual target, have great genetic adaptability, and the few hardy individuals surviving any chemical assault are able to incorporate resistance to their new chemical enemy into their offspring. By studying this phenomenon— as well as the process by which predators of target species often become pests themselves once their usual diet is temporarily reduced by chemicals—scientists have come to a much better understanding of how our agricultural ecosystems work. These insights are in turn leading scientists to new pest control strategies that could reduce or eliminate the use of pesticides altogether.

Furthermore, as it has become clear that the major classes of chemicals have already been exploited in the profitable search for pesticides,[2] the agrochemical manufacturers have shifted research money into a new arena. In the May 26, 1986, *Newsweek* Howard Schneiderman, Monsanto's senior vice president for research, was quoted as saying, "We'd like to wipe chemical pesticides off the face of the planet in 25 years."[3] What Schneiderman's statement refers to, however, is the current effort by Monsanto and other companies to create new life forms through genetic engineering, which will theoretically reduce or eliminate the need for chemicals to control pests.

The "Biorevolution," as it is called, has been slow to attract the public attention it warrants. Many people are only now beginning to find out about the old problem of pesticide residues in their food or water, in breast milk and in human fat tissue, and about the potential health risks those residues represent. With growing public anger focused on these serious questions, some companies are quietly abandoning their defense of pesticides and are moving toward a brave new world of genetic tinkering.

As it emerges from the private laboratories of the multinationals, the new biotechnology is entering a realm previously confined to science fiction. Using recently developed techniques, corporate scientists are now manipulating genetic material on the cellular and molecular level to create new organisms—not only "speeding up" nature, but doing things nature would perhaps never do on its own.[4] Ready for testing in the open environment—then to be rushed into commercial production—are exotic creatures like frost-fighting bacteria, one strain of which actually exists in nature already. In the test tube stage are cows as big as elephants, capable of producing 45,000 pounds of milk a year.[5]

No one doubts that the biorevolution carries with it the potential for major improvements in the human condition. But, as we should have already learned from our experiences with earlier promising technologies like petrochemicals and nuclear energy, the question of who controls technological innovation is ultimately more important than what, in an ideal world, that innovation might be able to contribute. One lesson the recent environmental disasters offer is that unanticipated side effects of technology *always* seem to occur in the "real world." Therefore, it is at the earliest stage of technological change—before massive investments create entrenched special interests—that we need careful review of the potential for future harm.

Measured by this standard, the current regulation of biotechnology is a disaster. Much more than was the case in the Green Revolution, for example, biotechnology research is "privatized"—under the control of the multinationals. As new life forms are created, altered, and patented, there are no effective national or international institutions overseeing the work. Regulations covering experimentation are vague and ineffective, and ominously absent in the Third World, where field experiments may well be conducted soon. (Many pesticides were—and are—tested in underdeveloped countries prior to registration in the developed countries.) What happens, for instance, if the mass-produced antifrost bacteria fail to encounter any predators in nature, and add to the potential threat (already represented by the hole in the ozone) to the polar icecaps? As improbable as this scenario seems, it is a fact that biotechnology raises global issues. Therefore decisions about its use should not be left to private corporations concerned mainly with maximizing returns for their shareholders.

Furthermore, at least one major thrust of the privatized research by the agrochemical companies casts substantial doubt onto the prediction by Monsanto's Schneiderman that biotechnology will replace chemicals. Over two dozen companies are developing herbicide-resistant crops, which if allowed to go on the market, will probably *increase* the use of herbicides in many places. This will happen because farmers will be able to treat their fields with chemicals that kill weeds but do not damage their crops.[6]

The biorevolution will create a whole new hazardous waste issue, as well, since cast-off genetic material will find its way into our sewers, rivers, soil, and air, further complicating any public efforts to control the spread of dangerous material in our common environment.

Under our modern food production system, as the

Bhopal Syndrome illustrates, some sacrifice of human life is actually deemed acceptable in order that a rich diversity of international food products reach those willing and able to pay for it. The main decisions on who will make that sacrifice are made by corporate managers, largely beyond public scrutiny. The average consumer buying a banana or a tomato at a shop in New York City or San Francisco has little way of knowing whose hands harvested those products days or weeks ago, thousands of miles away. On television, there may be a picture of "Juan," lovingly cradling coffee beans in a South American field, but there are seldom pictures of other Juans, dead from pesticide poisoning, in the coffee plantations where they worked.

With the advent of the biorevolution, private decision-making on how our food is grown greatly increases the risks to all of us. Global environmental risks will not discriminate in the long run between rich and poor, though the latter may initially bear the disproportionate share of the risk if there is a way to divide it. In determining the risks and benefits of technology in agriculture, an overall assessment of our *real* food needs is urgently needed. No publicly accountable institution currently conducts such an assessment.

For too long, the narrow boundaries of corporate self-interest, enlightened or otherwise, have confined the discussion of how each of us gets our food on the table. It is time now for that discussion to be brought out into the open, with free access to information and a strong commitment to the public's right to know. The current food production system took root in an era when our collective attention was elsewhere. Vast numbers of us have left the land during the past century to live in cities—a process that is only now reaching its peak in the Third World countries. In the process, we have left behind our traditional knowledge of how food is grown. In its place, we have entered into dependence on technologies that neither we nor the best human science fully understands.

The uncritical acceptance of these technologies has led to the current unsatisfactory state of affairs, where hundreds of thousands of people are poisoned by pesticides so that urban dwellers whose major nutritional problem is *over*-eating receive ever greater choices of foods. Pesticides and all other chemicals are regulated in a haphazard, inconsistent manner, with corporate lobbyists wielding more influence than those concerned about public health and environmental integrity.

For example, many chemicals, like daminozide—a plant growth regulator used primarily to cause apples to ripen simultaneously and thereby avoid the waste and rot of the natural growth cycle—become widely used before scientists have time to figure out how they work. In daminozide's case, the risk of cancer from a breakdown product that is created when apples are processed into juice or applesauce led to consumer protests and a remarkable "market response" by several large food distribution companies that recently refused to accept any more apples treated with daminozide. The chemical, whose toxic breakdown product is also a component of rocket fuel, had already been in wide use for 17 years, however, and in 1986 the EPA decided to allow it to remain on the market pending further study.[7]

What the daminozide case illustrates is that all too often we are willing to experiment with the public welfare—and particularly the health of children, who consume a large portion of the apple juice and applesauce produced each year—in order to financially benefit a handful of chemical producers and growers. The same risk-benefit equation holds true for many other pesticides, but the risks from biotechnology are even greater. If we allow artificial life forms into the environment, will they mate with existing creatures, joining the eternal dance of evolution of all life on earth? Given that the sum total of all the millennia of living and dying by all creatures, including humans, amounts to a thin layer of topsoil[8] (itself now seriously threatened by our agricultural and industrial practices), we should be much more modest and prudent before taking this next giant technological step.

In the absence of free and open public access to the information necessary to make intelligent decisions about pesticides and biotechnology, many of our regulatory measures end up based on faulty reasoning. We don't need more apples if it means exposing our children to cancer-causing substances in their applesauce. We don't need more oranges if it means releasing an artificial frost inhibitor into the environment. And we don't need more tomatoes in the winter if it means poisoning farm workers with pesticides in Mexico to get them.[9]

There are concrete steps that could be taken to improve the situation. Agricultural information could be added to the curriculum of all schools, including a program for urban high school or college students to work in farm fields for a semester or longer. An ecologically wiser agriculture would depend on more manual labor for such tasks as counting and gathering pests to determine when to intervene by releasing predators or employing chemical controls. A little time on farms would probably be exciting to many city-dwellers, and would build an informed constituency of citizens better able to understand both the difficulties of producing food, and the fundamentals of biology. The burgeoning community garden movement represents a similarly hopeful way toward these goals.

We also need a national commitment to agricultural research, with a new emphasis on ecologically sound techniques rather than the current bias toward chemicals and mechanization. And it will soon become apparent that we have lost much more than a quaint way of life with the destruction of small farms—a smaller-scale farming economy will be a necessity in the future, requiring new economic programs to regenerate what has been lost. There are, of course, many citizens' initiatives that have already recognized this need. The sustainable agriculture, land stewardship, regeneration, and permaculture movements are all part of a new agriculture that will eventually supplant today's monoculture as practiced by agribusiness. And the organic farming movement is finding a growing market for its produce, as more and more consumers exert the kind of pressure that will eventually erode the demand for chemically-treated food.

Finally, before reversal becomes impossible, our society will have to limit the concentration of wealth, power, and global decision-making by ever-larger, more tightly centralized, and undemocratic corporate empires. Whereas there are 20 or so major corporations deeply involved in the global pesticide business today, only half this number may be left five years from now.[10] There is such a thing as too big, especially in the case of control over the production and distribution of something as essential as food, and the agrochemical multinationals are clearly becoming too big for our collective good. It is because of their size and wealth and secrecy that the coming biorevolution could endanger us all. This final step will be by far the hardest to take, but it is the most critical in ridding ourselves of the Bhopal Syndrome, the disease of people captive to technology out of control.

Notes

1. R. H. Hall, "A New Approach to Pest Control in Canada," Canadian Environmental Advisory Council Report, no. 10, July 1981.

2. Ibid.

3. "Tinkering With Nature," *Newsweek*, May 26, 1986, p. 55.

4. Frederick Buttel, Martin Kenney, and Jack Kloppenburg, Jr., "From Green Revolution to Biorevolution: Some Observations on the Changing Technological Bases of Economic Transformation in the Third World," *Economic Development & Cultural Change*, vol. 34, no. 1, Oct. 1985, pp. 31–55.

5. Jack Doyle, *Altered Harvest: Agriculture, Genetics, and the Fate of the World's Food Supply* (New York: Viking Penguin, Inc., 1985), pp. 128–29.

6. "Tinkering With Nature."

7. Ibid.

8. Jane Thrall, Center for Investigative Reporting, May 1986, unpublished manuscript.

9. Joe Paddock, Nancy Paddock, and Carol Bly, *Soil and Survival: Land Stewardship and the Future of American Agriculture* (San Francisco: Sierra Club Books, 1986), p. 3.

10. Charles M. Benbrook, "Pesticide Regulatory Policy: Creating a Positive Climate for Innovation," a speech delivered at the Conference on Technical and Agricultural Policy, National Academy of Sciences, Washington, D.C., Dec. 12, 1986.

AFTERWORD

A Walk Through Bhopal

by Claude Alvares

> The children in the shantytowns of Bhopal have a new game. One plays the "mother," another the "father." Just as they have settled down for the night with their "children" around them, one shrieks, "Gas aagayi hai" (the gas has come!). Then they all leap up, thrash around, choke, and fall "dead."
>
> Vasanta Surya,
> journalist who visited Bhopal

The 22 Up Southern Express, originating in Delhi, touches Bhopal at nine in the morning, a journey of about 12 hours. One of the first sure indications that

EDITOR'S NOTE: Claude Alvares is a reporter and writer based in Goa, India, who specializes in environmental subjects. He has published widely on the problems of hazardous industries in Third World countries. This eyewitness report is based on Mr. Alvarez's four visits to Bhopal in June, July, August, and October 1985.

we have indeed reached Bhopal is the sight of the Union Carbide factory, at the periphery of the city. The plant is not easy to identify as such, for the signs and nameplates have been removed, as if in shame. And unlike other factories, this one is completely silent; waste gases no longer emanate from the numerous vents and chimneys.

Passengers retain the factory in their line of vision for quite a while: the rail track curves partly around it before entering the railway yard and then the Bhopal railway junction. The station is barely two kilometers from the Union Carbide plant. On the morning of December 3, 1984, all track activity here came to a stop for more than seven hours. Beggars sleeping on the platform never woke up. Soon, the railways had created a special hero out of the late station superintendent, Mr. Dhruwe, who is said to have persisted in his position, despite being choked by the gas, until he could telephone stations ahead of Bhopal to hold up all the trains. Once he had accomplished that, he is said to have forthwith collapsed and died.

Today, however, life at the station is back to normal. Unlike Hiroshima and Nagasaki, Bhopal did not suffer structural damage. So the appearance of normality has been easier to reconstitute, and it is now hard to imagine that the worst industrial disaster the world has ever known occurred here.

But what has happened to the gas victims who survived the tragedy? Have they gotten over it psychologically? Are toxins still embedded in their tissues, affecting their bodies? Is the medical and financial relief that's provided by the state government of Madhya Pradesh commensurate with the need? Will there be justice? Are there lessons here for all of us?

A large encampment of huts called J. P. Nagar, situated just outside the Union Carbide factory gate, is the community that bore the brunt of the descending gas that fateful night. The nearby plant is now in the hands of the Central Bureau of Investigation (CBI), and there is little possibility that it will ever be permitted to reopen. But the people it nearly scared out of existence are back: J. P. Nagar is once again a bustling slum, kept clean by its inhabitants, with dwellings constructed of wooden slats, most likely from used crates.

Outsiders are easily noticed here. There has been a continuous stream of journalists—Indian and foreign—TV crews, government officials, and prominent politicians. The inhabitants are quite poor, and they have learned that being victims of the gas tragedy at least brings them a few monetary benefits, which they could hardly have dreamed of before the accident. Certainly, before the lethal gas struck, they suffered continuous ill health due to poor, unhygienic water supplies, tuberculosis, and malnutrition. But these conditions do not justify special attention in India.

Thus, the moment it was known that I was a journalist, I became the confidant of a small crowd of *basti* (slum) dwellers, who insisted on accompanying me from house to house, and bombarded me with a series of complaints. The first house I visited had the number 0480 assigned to it by the Indian Council of Medical Research (ICMR), evidently as part of some ongoing medical investigation or monitoring related to the impact of the gas. I saw two children there. A six-month-old baby was blind in both eyes, with eyelids strangely glued together. A singing mass of flies pestered the child, as it lay motionless on the floor, undisturbed, unseeing, perhaps accustomed from birth to the insects. The older child had only one eye similarly affected. The father, Anip, brought me four empty boxes of Neosporin antibiotic eye ointment, which he had kept applying on the children's eyes, but with no relief. He had received four shots of sodium thiosulfate (NATS) as of June 1985, when the clinic was shut down by the police.

Usman Khan, another resident, told me he had lost his wife in the tragedy. Despite the fact that he has a receipt from the cremation ground authorities, a photo, and a death certificate from the municipality, he has received no compensation from the government. He has had to run around from official to official. He insisted that he had filled in all the necessary forms. His major problem was that his name did not figure in the initial survey made by the Bombay-based Tata Institute for Social Sciences.

Rafiq Khan said that he lost three members of his family, but was not present when the survey took place. The magistrate now wanted a bribe of 2,000 rupees to register each case.

Most of the complaints came from people who wished to establish that they had not yet received the ration cards for government doles; that they had not received the 1,500 rupees the government promised for all those affected by the gas; or that the names of their dead relatives were not included on the lists for compensation. One woman proudly showed me her photograph published in *Maya*, a Hindi magazine, in an article on the Bhopal disaster.

The similarity of the complaints is broken in House No. 342 by Ishaq, a small man, unshaven, wearing a white shirt over white pajamas. He recounted his grotesque tale with a seriousness that made us all laugh in spite of ourselves. Thought to be dead by officials after the accident, he was thrown into a truck, and only came to when the truck emptied its contents into the icy cold Narbada River. He managed to swim ashore. His case was widely reported in the papers, but thus far he has received no financial relief.

Rashid Khan (ICMR-417) lost his 12-year-old daughter, Mobina. No compensation, no free rations. The weary listing continued.

The Tata Institute is reputed to be one of the premier centers in the country for research, yet its survey of gas victims and fatalities in Bhopal is flawed because a sizable number of the victims have fled elsewhere, have not yet returned to the *basti*, or are in a hospital. From all indications, it is clear that no fresh surveys of gas victims are going to be entertained at the official level.

The inhabitants of J.P. Nagar have returned to their slum because at least there they own something. A year before the disaster, the state government, in one of its populist moves, had announced that henceforth all J.P. Nagar dwellers would own the land on which their houses stood.

But, due to their extreme poverty, these people are unable to travel elsewhere for better medical care. For the residents of the *basti*, recent health complaints are perceived merely as an additional burden to an already existing, long list of disabilities. Since the facilities for treatment of such standard diseases as tuberculosis are limited to ill-staffed public dispensaries manned by apparently apathetic doctors, it stands to reason that the management of gas-related disabilities would not be qualitatively different.

In India, the larger cities like Bombay and Delhi attract and retain the best medical personnel, and boast the most sophisticated medical equipment. The upper- and middle-class people who were affected by the gas in Bhopal naturally took off for these places, by train, bus, and airplane. All hoped that personal health care would improve their chances of escaping the effects of the gas. That hope proved to be unfounded, however. There are few doctors even in these cities who know how to manage such symptoms with any level of competence. The poor *basti* dwellers of Bhopal would be consoled to hear that even the rich have no other prospect but to bear with them the involuntary penance of being permanently disabled by the gas in one way or another, that all the wealth in the world would not enable anyone to breach the barriers of medical ignorance in this case.

Anthony Thomas Henriques, an employee of the Greaves Cotton Company in Bombay, was in Bhopal on company business on December 3, 1984, when, he told *Society* magazine, he rushed out of the hotel to see death everywhere.

"Oh my God! Why are all these people lying on Hamidia Road? They were dead, every one of them, hundreds of them—dead. They were all trying to get away from the gas. The dying were urinating, defecating, and vomiting. Instead of running on the main street, I ran into the bylanes and through the hutment area. I was choking because the gas was around. I stopped at a hut and asked a weeping woman to give me some water.

" 'What has happened?' I asked her.

" 'It's gas,' she said.

" 'You will die here,' I said.

" 'All are dead,' she replied. 'I will also die here.'

"I ran to the lake—the lake which leads up to the station going to T.T. Nagar, New Market. There is a small temple here beside the lake. I ran down—Oh, my God, thousands of people were lying there. Some were alive, they were drinking water, they were vomiting in the lake and drinking the same thing. Many were dead, choked, dying bodies were heaped up . . . one dying on top of another. . . . I lay down beside the dead bodies and dipped myself into the water. I stripped my T-shirt off me and held the wet shirt in front of my face.

"In a few minutes I will be dead. What will happen to my body? It will rot here, or someone may burn my remains in this chamber of death. Back home, my wife, children, and parents must be dead. This is a nuclear explosion, it cannot be anything else. . . . This is the end of the world. I'm dying. . . .

"After some time I picked myself up and started running again. Suddenly I found myself on the police grounds. I was finished. There is a neem tree there. I put my arms around it and started weeping and praying. As I hit the ground, my body touched something. It was a dead body, looking terrible. . . . Oh, my God, it gave me such a jolt, I unconsciously began tumbling away, then I began running again. I was seeing gas everywhere, because my eyes and mind were affected. . . . I was imagining it. . . .

"So many dead, so many dying. During this time, there was no love left in Bhopal. Husbands were leaving their wives, and mothers were leaving their children and running for their lives. People who had been calm all their lives were screaming at one another, filled with fear and hate, people were dragging one another . . . it was hell. . . . Finally I managed to get out of Bhopal. During this time it was like the aftermath of a nuclear war. If ever there is a hell, it must look like Bhopal on that gas day. It looked like the end of the world."

In Hiroshima and Nagasaki, in 1945, the kind of medical problems created by the atomic blast posed grim, unprecedented challenges. Japanese society, recognizing that it was faced with a problem that would extend itself over several generations, put research and relief on a coordinated, long-term basis. Even today, 40 years after the blast, the names of victims who die from radioactivity-related illnesses are religiously added to a long list preserved in the memorial to the bomb victims. Living victims make two visits a year to specially designated health centers where their condition is continuously monitored.

In that respect, Bhopal represents the reverse of Japan. Here, a majority of the gas victims were not only poor, they belonged to a minority community. Over and above these hard facts of Indian life, already formidable in themselves, hovered the perpetrator, a multinational corporation that had managed over the years to befriend and benefit the ruling elite of the city, including the medical profession. The trajectory of the aftermath, of prospects for relief, should have been fairly easy to predict in such circumstances. The factory had operated without safeguards because of the socio-political matrix in which it was located. It follows that the management of relief, of rehabilitation, would in no realistic sense operate outside the same matrix.

Within three days of the disaster, it became apparent that besides MIC, more than 20,000 pounds of dangerous breakdown compounds like hydrogen cyanide and the cyanogen chlorides had been possibly also released from tank No. 610 during the accident. Nevertheless, a sizable and influential number of scientists and doctors in Bhopal publicly asserted that no cyanide deaths had taken place. The antidote to cyanide poisoning is sodium thiosulfate (NATS), but literally thousands of people were given secondary treatment including large doses of dangerous corticosteroids when they could have been given NATS instead. The medical authorities even issued a circular (dated December 13, 1984) banning the use of the drug.

It was only in early February that the ICMR issued guidelines for the use of NATS on the basis of a double-blind trial of the drug. A month later, the director of health services was still telling the press that he had not received any such communication from the ICMR. And until June 1985 the state authorities did all they could to prevent NATS therapy from being administered routinely to the gas victims.

The mysterious sabotage of the NATS campaign was closely orchestrated with an official propaganda blitz, in which the damage from the gas leak was proposed to be less severe than what had been thought earlier. The victims were said to be undergoing mostly mental stress, and to be more in need of psychiatric aid than medical care. This gradual shift in the official perception of the nature of gas-induced disability from a physical to a psychological basis would also imply that the tragedy henceforth would be only indirectly connected with Union Carbide. The state government also indicated that a large number of the gas victims were basically suffering from a revival of their earlier illnesses, particularly tuberculosis, contracted much before the gas leak, and therefore presumably unrelated to questions of accountability or aid.

A doctor from Hamidia Hospital contradicted this position, however, in late July:

"The incidence of tuberculosis has gone up 20 percent. Those cases that were very nearly cured have recommenced: the severe coughing fits brought on by inhalation of the gas have eroded the fibrosis. Localized infections have become widespread. The incidence of tuberculosis in the area was already quite high due to the presence of a textile mill. Asthma cases have also deteriorated: those with chronic asthma histories find that prescribed drugs have less effect than before. Borderline asthma cases have become full-blown. There has been an increase in chronic obstructive-airway diseases. The oxygen-carrying capacity of the blood has been reduced."

"My name is Kailash Pawar. I am a gas victim. I am 25 years old and I work as an auto-rickshaw driver. I was rendered unconscious by the gas, and they put me in a truck with corpses destined for the cremation ground. I awoke with the impact, when I was thrown from the truck onto a heap of dead bodies. My wife died. I have a four-month-old girl, who is not well at all.

"I have received 10,000 rupees as compensation towards the death of my wife, but all this has been exhausted in funeral costs, medicines, and other expenses.

"I was hospitalized on 5th December, but was told I would have to pay 3,000 rupees if I wished to occupy the bed. I had to leave. On 13th January my condition deteriorated. I was admitted to the Kasturba Hospital. I became unconscious. On the 20th, doctors started me on a new drug [sodium thiosulfate.] The very next day I was already better. The oxygen was stopped, and soon all my symptoms disappeared. I was discharged on 30th January.

"Since I was in hospital at the time the survey of the gas victims was done, I was not included in the list. In the meantime, the lower division clerk in charge of my area had been issuing my rations to himself. My applications for free rations were denied. It was only on 4th July, after I met the collector, that I was finally issued free rations.

"The symptoms of breathlessness began to recur ten days ago. This time when I went again to the hospital, they refused to admit me, for they said I was not a gas victim. I had to get this drug [sodium thiosulfate] from a private doctor before I got relief."

Within days of the gas tragedy, in December 1984, a major citizen's group formed called the Zahreeli Gas Sangarsh Morcha (Poisonous Gas Event

Struggle Movement). Led by radical scientists, the group initially focused on the rights of the victims to proper information concerning the factory, the cause of the accident, and the effect of the gas on the victims' long-term health.

Within a few weeks, differences had arisen within the group, and a splinter organization was formed— the Nagrik Rahat Aur Punarvas Samiti (Citizens' Relief and Rehabilitation Committee). Both organizations now vied with each other to represent the cause of the victims.

The Morcha was responsible for a national conference in February 1985, at which more than 150 people from all over India, including scientists, doctors, and members of civil rights groups, got together to discuss the Bhopal disaster, its causes and aftermath. The Morcha also attempted to coordinate a nationwide signature campaign directed to Prime Minister Rajiv Gandhi, demanding among other things that the Union Carbide factory be taken over and converted into a hospital. While the Punarvas Samiti concentrated on direct relief in a few small pockets, the Morcha moved in the direction of a confrontation with the local state government on various issues.

Eventually, in May 1985, the Morcha and the Samiti, together with Union Carbide's Workers' Union and the Trade Union Relief Fund of Bombay, cooperated to form the Janasasthya's (People's Health)

Samiti, and opened a clinic where victims would be administered the antidote NATS. The clinic was within the Union Carbide factory premises. By the end of June, the agitational politics of the Morcha reached a new peak when about 2,000 gas victims and the police clashed outside the state secretariat.

The state government seems to have made a firm decision to silence the Morcha and any activities connected with it. Accordingly, not only were Morcha activists arrested and charged with criminal offenses, but the doctors running the clinic were arrested as well. The clinic was shut down and the medical records handed over to Union Carbide's security officer. It was at a time like this that the Morcha's shaky links with Bhopal's unaffected population became apparent: It was unable to find people to stand security for bail for those arrested from within the city without considerable difficulty. Eventually, it had to approach the Supreme Court in Delhi to enforce its right to reopen the clinic.

"My name is Ashay Chitre. I am a gas victim. So is my wife, Rohini, and our five-month-old son, Yohul. I got up at three o'clock that Monday morning, 3rd December. It was dark. The sound of moaning, crying, coughing faded in. . . . There was traffic, unusual for the city to be up this early.

"I got up, switched on the light, and approached

the window. The sound grew, it sounded like a child crying softly, only it was amplified. I parted the curtain, and felt a blast, something invisible entered the room. My eyes started smarting, watering. I wanted air. . . . As I turned around to face a blurred image of my wife—I knew that the sound outside was our own—her eyes were streaming and she was breathless. . . .

"Rohini grabbed a bedsheet and a pair of sunglasses. I told her it would probably be a long walk. . . .

"The air outside was heavy, we fought to breathe; fluid streamed out of our eyes and nose, and energy was sucked out of us. There was chaos outside, people were emerging from all directions. A few hundred yards later, Rohini (six-months pregnant) had a coughing fit and collapsed. I fought to keep my eyes open and said, 'Think of the baby, don't give up.' Somehow I managed to get her onto her feet. People fled past us, coughing and gasping for air, some tried to cling onto speeding, overflowing vehicles—cycles, rickshaws, cars, trucks, even handcarts. . . . Some were vomiting, some just collapsed and we had to walk over them—not knowing whether they were dead or alive.

"From the dark bylane we entered the sodium light *chowk*—there were hundreds of people all over—the light cast a monotone sepia color over the scene, like an old photograph—in it we saw our own past. . . .

"Families had piled their few belongings and

many children onto handcarts; some sat huddled together, giving up by saying, '. . . if we die, let us die together. . . .' Others left the very young and very old behind, like memories forgotten. As death filled our expanding chests we all became one, not knowing what we were running away from, or where.

"Feeling safer with people around me. . . . I ran, walked, stumbled to the nearest telephone. . . . I fell thrice while dialing but somehow got through to a friend in New Bhopal [which is situated on a height]. I went back to Rohini and we stood close to the trees where we found some relief. . . . It was 5 A.M. when my friends picked us up. The police were arriving and telling people to go back home; it was all over they said. Our eyes were still smarting. I asked, what was over? And, even today, don't get a reply."

An additional chief secretary of the Madhya Pradesh Government, Dr. Ishwar Das, was appointed relief commissioner a full five months after the gas disaster. Dr. Das is a senior bureaucrat, more comfortable with statistics, perhaps, than with people. He was in the United States when the gas enveloped Bhopal, and did not return till a full two weeks after the disaster.

In an interview, he told me he thought the main lingering health problem was lung damage. At least 10,000 of the victims were "burnt-out cases," with passages permanently damaged, and with acute breathlessness. He tended to minimize every other problem except this. During the entire interview, he made no mention of a possible "cyanogen pool" continuing to exist within victims' bodies and requiring detoxification in some form or another. The Indian Council of Medical Research had found this the most significant empirical finding of its ongoing research. Dr. Das admitted that NATS provided "subjective and symptomatic relief," but was reluctant to discuss this subject.

Dr. Das admitted gastroenteritis cases, cases with damaged esophagus and stomach, ocular problems in the form of photophobia, and watering of the eyes. But he was emphatic that no cases of blindness were discovered, even while conceding that there was permanent impairment of eye function and considerably reduced vision in thousands of people. He also denied that there would be any teratogenic effects among newborn babies (an opinion that would be proved wrong later, when babies conceived in the first three months after the gas leak would enter the world dead, eyes shut, lungs collapsed).

Dr. Das also provided some other curious information. He observed that out of a population of 800,000 people in Bhopal, some 600,000 were being given free wheat and rice, though the affected population was not more than 250,000. The dole cost the state government 2 crores (20 million rupees) a month. In the name of gas relief, some 300,000 af- fluent people from non-gas-affected areas were getting such free supplies, while bureaucrats and clerks con- tinued to cheat genuine but poor gas victims of these very same rations. The decision to place the entire city on a food dole was made by the chief minister just before the state elections, which, in Bhopal City, had been postponed until January 1985 because of the accident.

I was unable to fathom why Dr. Ishwar Das was so keen to *minimize* the impact of the gas disaster, and I left his office strangely perturbed. He had sought to create the impression that the relief operation—over- seeing the health status of some 250,000 affected people—was well in hand. Yet an elementary calcu- lation from published government data clearly proved otherwise. For example, the government had conced- ed that there were still 10,000 cases of drastic pul- monary damage, yet all it could concretely show was the reservation of 60 beds in Hamidia Hospital and 30 in the DIG Bungalow for gas victims in general. Clearly the scope of the medical relief hardly matched reality.

I visited the special 30-bed hospital, which is 100 meters from the Union Carbide factory. I was ex- pecting a special ward, with 30 beds, kept exclusively for gas victims. Instead, I discovered a general ward, in which gas victims must be painstakingly located from among others suffering from diarrhea, tuberculosis, and other illnesses. A cursory glance at the patients' records indicates that thiocyanate levels in urine are being monitored. Beyond that, nothing very special was available for their treatment. After I spoke with some of the victims, I asked for a discussion with the superintendent. She refused to talk to me, saying only that instructions had come from the highest health officials not to discuss the cases with anyone without prior permission from the top.

Neither is one permitted any access to medical records. The secrecy surrounding them borders on the hysterical; early newspaper reports had even indicated large-scale burning of medical records of the first days following the disaster.

Thus, it would seem that those with Union Carbide's interest in mind had succeeded in containing any possible anguish about real, long-term damage to human beings. The Bhopal government's officials do the same job with quiet sophistication; a TV documentary and a six-page advertisement in leading Indian magazines proclaim how well the state government is looking after the "poor gas victims." And despite the number of gas-related deaths that have occurred during every single month of 1985, the earliest announced figure of people dead—1,754—has never been adjusted.

Soon, the Bhopal tragedy may even conceivably

cease to exist as such. Officially, the charade of relief and rehabilitation will continue, for the central government needs legitimacy within the American courts. But in Bhopal itself, one could as well say the show is being wound up, with government officials suggesting that gas-related health problems exist predominantly in victims' minds, and that most cash payments have by and large been met.

> I don't think we would just like to be around and pretend we are sick. We too want to do something in life. We also want to work. We have a baby. We want to look after him. Because we are not the kind of people who would like to sit at home and who would just like to claim compensation. We want to do something but the body is refusing to let us do anything. That's the sort of thing that we are trying to overcome but it's really hard.
>
> Rohini Chitre, gas victim

The Punjab Mail leaves Bhopal for Bombay at 8 P.M. That evening, before I boarded, I was told by a wealthy lady, who was unaffected by gas since she resided in New Bhopal where the elite live, that though these poor *basti* folk had indeed suffered, they should cease to agitate so much, since "the government was doing so much for them," including giving them all free rations of wheat and rice. Earlier that week, radical groups like the Morcha and the doctors at the

People's Health Clinic had been meted out unbending oppression from the state government. Newspapers had begun referring to the doctors running the clinic as "outsiders," since they had come from Bombay and Calcutta.

Bhopal is a government city—almost 70 percent of the city's population is in some form of government service. We can thus imagine a new kind of scenario. Because the government is the source of income for the majority of the city's population, it is enabled to call the shots. It can decide, directly or otherwise, that victims of the gas disaster need no longer be recognized as such, that doctors do not agree to register or identify people as gas victims. It can launder or destroy medical records, it can apportion compensation money as it wishes, not as lump-sum payments to the relatives of dead persons, but to socially more "useful" purposes: to develop the city, rebuild its roads, and generally collaborate with Union Carbide in the latter's plans to improve its image in Bhopal.

Since the general tendency in India for the past 40 years has been in the direction of getting more and more people dependent for their livelihood on the government, Bhopal could provide a model of how to cope with such major disasters in the future. Cleaning up rapidly after such major accidents is all-important, in so far as these disasters have unwanted potential to distract societies from the hazardous path of economic development thrust on them.

Government servants do not, or generally will not, question government orders. They are generally a meek folk, worried about job security first. If the government should order that any serious disabilities that continue to be registered are not due to gas, but to other causes, so be it. Nobody wants to remain a victim. If being a victim is politically undesirable, well, the entire city could conceivably collude in such a cover-up. In that sense, ringing down the curtain on the Bhopal disaster within a year is itself a major strategic success.

But then, whatever happened to all the pain? Can that all be forgotten? How many hundreds sleep uneasily, and will do so for the rest of their lives, dreading the prospect of drowning in another chamber of gas, with no premonition, no route out of its evil and terrifying influence, eyes blinded, lungs clogged, the agony ending only when the gas arbitrarily disappears? What of children lost or children orphaned? People died in pain, they suffered unreasonably for days and months, and continue to suffer. In America, such anguish by itself would constitute sufficient grounds for a severe compensation suit.

Why does the wealthy lady criticize victims and their families? Why do gas victims fear to be left alone any longer in their houses in Bhopal? What is the kind of consolation one offers mothers delivering premature or diseased infants?

Governments do not understand compassion or

pain. They prefer to deal with statistics, schemes for relief, dollars, or doles. If people do not recover after all the aid the government feels it has so magnanimously distributed, then something is wrong with the people. Government officials claim they are men in a hurry, and that nothing should be permitted to discourage the welcome aid of progress. Suffering can be managed and normality assured, at least on the surface, for all the world to see. Disasters are a nuisance; Bhopals occur every day, in a thousand different ways. Nobody makes a fuss. It is accepted, whether we like it or not, that suffering and death are a permanent feature of all modern progress.

Throughout history, terrible disasters have often spawned mythologies that capture the total helplessness of the victims. In the *bastis* of Bhopal today, an old woman is said to visit houses unannounced. She knocks at the door and asks for food. If she is given it, she throws it at the house, and all the inhabitants die. If she is refused food, she curses those who live there, and all of them will die.

The newspapers report that recently, in the old part of Bhopal, frightened residents joined together to take part in a beating. Their victim was an old woman.

Appendices

Appendix A

Top 20 World
Agrochemical Producers
1983

Company (home country)	Sales (millions of dollars)
Bayer (FRG)	1,500
Ciba-Geigy (Switz.)	1,320
Monsanto (U.S.)	1,167
Shell (U.K./Neth.)	720
ICI (U.K.)	695
Rhone-Poulenc (France)	630
Hoechst (FRG)	620
DuPont (U.S.)	580
BASF (FRG)	568
Dow (U.S.)	565
Schering (FRG)	435
Eli Lilly (U.S.)	378
FMC (U.S.)	340
Rohm & Haas (U.S.)	337
Union Carbide (U.S.)	335
Stauffer (U.S.)	325
American Cyanamid (U.S.)	255
Kumiai (Japan)	254
Chevron (U.S.)	240
Sandoz (Switz.)	210

Source: Wood, Mackenzie & Co. Agrochemical Service, 1984

Appendix B

Major Pesticide Intermediates
1980 (est.)

(Percentage of World Market)

Phosphorous compounds	10.7%
Amines	7.9%
Cyanuric chloride	6.6%
Phenol derivatives	6.0%
Aniline derivatives	5.7%
Mercaptans	5.1%
Phosgene derivatives	3.6% (includes MIC)
Chloroacetyl chloride	3.4%
Carbon disulfide	3.0%
p-Chlorobenzotrifluoride	2.2%
Monochloroacetic acid	2.1%
Maleic anhydride	1.6%
Cyclodienes	1.6%
Pyridines	1.6%
Chloral	1.0%
Benzoic acid derivatives	0.7%
Guanidines	0.4%
m-Phenoxybenzaldehyde	0.1%
DV acid	0.06%

(Note: These percentages are estimates based on an estimated world market of $2.5 billion for intermediates in 1980, not counting the USSR, Eastern Europe, and China. Although the above figures are approximations, and should not be considered as precise, they are an indication of the extreme fragmentation of the intermediate market, particularly since the above list accounts for only about two-thirds of the total market.)

(Compiled from confidential industry sources.)

Appendix C

Major Industrial Accidents Over the Past Seven Decades

The MIC gas leak in Bhopal, India, is generally regarded as the worst industrial accident in history, both in terms of injuries and loss of life. The following list is a compilation of the major industrial accidents that have occurred over the past 70 years. Accidents included in the list have either caused a large number of deaths and/or injuries, or involved a major evacuation of people. The list documents accidents stemming from the transportation of hazardous substances, as well as chemical spills, and explosions at industrial facilities.

- December 6, 1917,
  Halifax, Nova Scotia, Canada:

A French ship carrying about 1,000 tons of ammunition collided with a Belgian steamship, setting off explosions that destroyed a two-square-mile area of Halifax and damaged nearby piers. Some 1,654 people were killed.

- September 21, 1921, Oppau, Germany:

The biggest chemical explosion in German history occurred at a nitrate manufacturing plant about 50 miles south of Frankfurt. The blast destroyed the plant, a warehouse, and leveled houses four miles away in the nearby village of Oppau. At least 561 lives were lost and some 1,500 people injured.

- October 20, 1944, Cleveland, Ohio:

A poorly designed liquefied natural gas tank belonging to the East Ohio Gas Company developed structural weakness, resulting in a massive explosion. The ensuing blast and fire claimed some 131 lives.

- April 16, 1947, Texas City, Texas:

A freighter, the "Grand Camp," carrying 1,400 tons of ammonium nitrate fertilizer exploded after fire broke out on board. The initial explosion set off a series of secondary explosions that destroyed much of Texas City. The blast rattled windows 150 miles away and the leaping flames also destroyed a nearby Monsanto factory producing a combustible ingredient of synthetic rubber, i.e., styrene. The next day another freighter, the "High Flyer," also loaded with nitrates, exploded in the same harbor. Some 576 people were killed and 2,000 others seriously injured.

- July 28, 1948,
  Ludwigshafen, Germany:

A railroad car transporting dimethylether (used in the manufacture of acetic acid and dimethylsulfate) to the I.G. Farben chemical plant exploded inside the factory gates. The blast and resulting fire killed 207 people and injured 4,000 others.

- August 7, 1956, Cali, Colombia:

Seven trucks loaded with dynamite exploded in the center of Cali, destroying over 200 buildings and leaving a crater 85 feet deep and 200 feet wide. Some 1,100 lives were lost.

- June 1, 1974, Flixborough, UK:

A railroad car carrying cyclohexane at a NYPRO Ltd. plant ruptured, resulting in the escape of about 400 metric tons of the chemical. The cyclohexane cloud exploded, setting off a fire over 20 acres of land. The blast killed 28 people, injured 89, and 3,000 others were evacuated; the blast also leveled every building on the 60-acre plant site.

- July 10, 1976, Seveso, Italy:

An uncontrolled exothermic reaction in a reactor at the Hoffman-La Roche Givaudan chemical plant caused a terrific explosion. The ensuing release of some 10–22 lbs. of toxic tetrachlorodibenso-p-dioxin contaminated soil and vegetation over 4,450 acres of land, and killed over 100,000 grazing animals. Although there were no immediate injuries or loss of human lives, over 1,000 residents were forced to flee, and many children subsequently developed a disfiguring rash called chloracne.

- July 11, 1978,
  San Carlos de la Rapita, Spain:

An overloaded 38-ton truck carrying 1,518 cu. ft. of combustible propylene gas skidded around a bend and slammed into a wall, sending 100-foot-high flames into a campsite where 780 tourists were eating, sunbathing, and swimming. Some 215 lives were lost and many people were injured.

- April 8, 1979, Crestview, Florida:

Seventeen railroad cars carrying acetone, anhydrous ammonia, carbolic acid, chlorine, and methanol derailed. The chemicals released from the ruptured railroad cars triggered fires and explosions. No one was killed, but 1,000 people were injured and 4,500 others evacuated.

- November 10, 1979,
  Mississauga, Ontario, Canada:

A total of 21 railroad cars carrying caustic soda, chlorine, propane, styrene, and toluene derailed. Three of the railroad cars carrying propane and toluene exploded and caught fire while a fourth railroad car carrying chlorine ruptured and a fire ravaged its contents without exploding. No lives were lost, but eight fire fighters were injured and 250,000 people evacuated.

- April 3, 1980,
  Somerville, Massachusetts:

A railroad car carrying phosphorous trichloride collided with a locomotive and spilled 6,000 gallons of the chemical. No one was killed, but 418 people were injured and 23,000 others evacuated.

- June 5, 1980, Port Kelang, Malaysia:

A fire resulted in the explosion of cylinders containing ammonia and oxyacetylene. The explosion, heard 15 miles away, caused extensive damage and the port had to be closed for six weeks. Three lives were lost, 200 injured, and over 3,000 people evacuated.

- June 6, 1980, Garland, Texas:

Nine railroad cars derailed and ruptured, spilling 5,000 gallons of styrene. No lives were lost, but five people were injured and 8,600 others evacuated.

- July 26, 1980, Muldraugh, Kentucky:

The derailment of 18 railroad cars caused two railroad cars carrying vinyl chloride to rupture and go up in flames. No one was killed, but four people were seriously injured and 6,500 others evacuated.

- July 27, 1980, Newark, New Jersey:

A railroad car ruptured and caught fire, releasing 26,000 gallons of ethylene oxide. No lives were lost, but 4,000 workers were evacuated, along with people within a half-mile radius of the site of the accident.

- May 19, 1981,
  San Juan, Puerto Rico:

A valve failure at a chemical plant released two tons of chlorine. No one was killed, but 200 people were injured and 2,000 others evacuated.

- August 1, 1981, Montana, Mexico:

A train derailment resulted in the rupture of two railroad cars carrying chlorine, releasing 90 tons of the chemical. At least 29 people were killed, over 1,000 injured, and 5,000 others evacuated.

- April 10, 1982, Belle, West Virginia:

A pipeline carrying chlorine gas burst, releasing 29 tons of the chemical. No one was killed, but 13 people were injured and 1,700 others evacuated.

- September 28, 1982, Livingston,
  Louisiana:

The derailment of 43 railroad cars carrying styrene, toluene diisocyanate, vinyl chloride, and other chemicals, resulted in spills, explosions, and fires lasting seven days. Although no lives were lost, over 2,800 people were evacuated.

- April 3, 1983, Denver, Colorado:

The puncturing of a railroad car, during switching operations, released 20,000 gallons of nitric acid. No one was killed, but at least 43 people were injured and 2,000 others evacuated.

- February 25, 1984, Cubatão, Brazil:

Gasoline leaking from a pipeline exploded, setting off fires in a shantytown in the Brazilian state of Sao Paulo. At least 500 people were killed and many others seriously injured.

- November 19, 1984,
  Mexico City, Mexico:

Shortly before dawn, some 80,000 barrels of liquefied natural gas tanks exploded at the San Juan Lxhuatepec storage facility operated by state-owned Petroleos Mexicanos. The resulting fire claimed 452 lives and injured 4,248 others; 31,000 people were evacuated and 1,000 others are reported to be still missing.

- December 2–3, 1984, Bhopal, India:

The escape of some 40 tons of MIC gas from a Union Carbide pesticide production plant in the Indian city of Bhopal led to the world's worst industrial disaster. At least 2,500 people were killed, 10,000 seriously injured, 20,000 partially disabled, and 180,000 others adversely affected in one way or another; some 150,000 people are reported to be still suffering from the adverse effects of the Bhopal catastrophe.

- January 6, 1985,
  Koratty, Kerala, India:

Toxic fumes of chlorine leaked from the dyeing section of a large privately owned textile mill. Although no lives were lost, some 40 workers had to be hospitalized.

- January 26, 1985, Cubatão, Brazil:

A pipeline burst at a government-owned fertilizer manufacturing plant and spewed out 15 tons of liquid ammonia. No one was killed, but over 400 local residents were administered oxygen at first-aid posts and some 5,000 others evacuated.

- August 12, 1985,
  Institute, West Virginia:

A 300-meter long cloud of toxic gases leaked from a Union Carbide pesticide production facility that had been redesigned, with the installation of a $5 million safety system, in the wake of the Bhopal tragedy. Although there was no loss of life, some 135 people, including six workers at the plant, were overcome by breathing problems, burning sensation in the eyes, and nausea.

- April-May, 1986, Chernobyl, USSR:

An explosion at a nuclear reactor spewed radioactivity over a wide area of the Ukraine, killing an uncertain number of people, damaging cropland, and contaminating food supplies in many European countries. Trace amounts of radioactivity were measured all over the world in the wake of the disaster.

- October-November, 1986,
  Basel, Switzerland:

A fire at a chemical warehouse owned by Sandoz, Inc., sent massive amounts of toxic chemicals, including 66,000 pounds of pesticides, into the Rhine River. Ciba-Geigy also dumped a smaller amount of pesticides into the Rhine during the same time period.

(Source: Martin Abraham, *The Lessons of Bhopal: A Community Action Resource Manual on Hazardous Technologies*, IOCU, Penang, Malaysia, September 1985.)

Appendix D

U.S. Chemical Accidents
(1980–85, partial list)

Total Accidents	6,928
Deaths	139
Injuries	1,478
Evacuations	217,457
Toxics Released	420 million lbs.

(Note: This list is incomplete, having been compiled from a variety of sources, both official and unofficial, including press clippings. The actual totals for accidents, injuries, evacuations, and toxics released is apparently much higher.)

Source: U.S. Environmental Protection Agency Hazardous Events Database

Appendix E

Sale of Union Carbide's Assets

Since the accident at its Bhopal plant, Union Carbide has sold off many of its assets, and is now a much smaller company than it was at the time of the tragedy. The proceeds from these sales have mainly been distributed as payments to shareholders, raising the possibility that the company will declare bankruptcy when its liability is eventually determined by the courts. The following table lists the sales and divestitures of Union Carbide since the disaster. (Sales price, where known, is listed in parentheses.)

March 1985:   Sale of cryogenic equipment manufacturing business to Harsco Corp

July 1985:   Sale of domestic welding and cutting system business to L-Tec Co. ($57 million)

Oct. 1985:   Sale of plastic grocery bag and drinking straw business to Jet Plastica Industries Inc.

Nov. 1985:   Sale of Canadian and West German welding systems businesses to local affiliates of L-Tec Co. ($13 million)

Dec. 1985:   Sale of films packaging business to a subsidiary of Envirodyne Industries Inc. ($230 million)

Dec. 1985:   Sale of engineering polymers and composites business to Amoco Chemicals Corp. ($210 million)

Dec. 1985:   Sale of Westchester, N.Y., lab and office building and 275 acres to Keren Ltd. Partnership ($170 million)

Dec. 1985:   Sale of South African chromium business to General Mining Union Corp; tungsten and vanadium business to an employee group from Umetco Minerals Corp., a Union Carbide subsidiary ($83 million)

April 1986:   Sale of home and auto products business to First Boston Inc. ($800 million)

April 1986:   Sale of battery products division in United States and some foreign countries to Ralston-Purina Co. ($1.4 billion)

Sept. 1986:   Sale of assets of Brownsville, Texas, chemical plant to RIO Systems Inc. ($10 million)

Nov. 1986:   Sale of agricultural chemicals business to the Rhone-Poulenc Group ($575 million)

Nov. 1986: Sale of Danbury, Connecticut, headquarters building and 650 acres of land to Related Companies ($340 million)

Nov. 1986: Sale of assets of electrical carbon business to Morgan Crucible Co. ($25 million)

April 1987: Sale of 50 percent of its electronic capacitor business, which has annual sales of $200 million and 6,000 employees, to senior management, financed by General Electric Credit Corp., which has the right to acquire a 35 percent interest in the new operation ($150 million)

Sources: *Business Week*, *New York Times*, *Wall Street Journal*

Index